Water and Climate Change in Africa

In the coming decades, countries around the world will face increasingly severe challenges related to global climate change. While the details vary from country to country, the impacts will be especially grave for marginalized people, whose access to food, potable water, and safe shelter may be threatened due to fluctuations in rainfall and temperature, and to extreme weather events. Because weather extremes are the main way that climate change manifests itself, water governance is a crucial aspect of climate change resilience.

Following an overview of the ways climate change is affecting three cities in Africa, *Water and Climate Change in Africa: Challenges and community initiatives in Durban, Maputo and Nairobi* discusses the equity and climate justice implications, and then gives examples of ways in which a range of local community organizations are extending their current activities to address these challenges through innovative new programmes and initiatives at the grassroots. This approach has implications for communities worldwide; it is a process of building on existing organizations' aptitudes and strengths in the light of local knowledge of climate challenges, and creating partnerships to build equity-enhancing new methods of protecting people's subsistence.

This book should be of interest to climate change scholars, activists and policy-makers, as well as development studies researchers and practitioners.

Patricia E. Perkins is a Professor in the Faculty of Environmental Studies, York University, Toronto, Canada.

Routledge Advances in Climate Change Research

Local Climate Change and Society
Edited by M. A. Mohamed Salih

Water and Climate Change in Africa
Challenges and community initiatives in Durban, Maputo and Nairobi
Edited by Patricia E. Perkins

Water and Climate Change in Africa

Challenges and community initiatives in Durban, Maputo and Nairobi

Edited by
Patricia E. Perkins

Routledge
Taylor & Francis Group
LONDON AND NEW YORK

First published 2013
by Routledge
2 Park Square, Milton Park, Abingdon, Oxon OX14 4RN

Simultaneously published in the USA and Canada
by Routledge
711 Third Avenue, New York, NY 10017

Routledge is an imprint of the Taylor & Francis Group, an informa business

British Library Cataloguing in Publication Data
A catalogue record for this book is available from the British Library

Library of Congress Cataloging-in-Publication Data
Water and climate change in Africa : challenges and community initiatives in Durban, Maputo and Nairobi / edited by Patricia E. Perkins
p. cm. — (Routledge advances in climate change research ; 2)
Includes bibliographical references and index.
1. Climatic changes—Africa. 2. Climate changes—Economic aspects—Africa. 3. Climatic changes—Environmental aspects—Africa. 4. Water-supply—Africa—Management. 5. Water resources development—Africa. 6. Water-supply—Government policy—Africa. I. Perkins, Patricia E. II. Series: Routldege advances in climate change research ; 2.
QC903.2.A35W38 2013
363.73874096—dc23

2012043536

ISBN: 978–0–415–82203–9 (hbk)
ISBN: 978–0–203–55604–7 (ebk)

Typeset in Sabon
by RefineCatch Limited, Bungay, Suffolk

Contents

Figures

Boxes

Contributors

Sadique Bilal Issa is Sports and Sports Facilities Coordinator for the Kilimanjaro Initiative, based in Nairobi, Kenya. Kilimanjaro Initiative's mission is "To encourage young people to have self-belief and to assist in providing opportunities that will enable them to take on a constructive role in their communities."

Patrick Bond is Senior Professor and Director of the Centre for Civil Society at the University of KwaZulu-Natal, Durban, South Africa, as well as a climate justice activist and political economist for social and labour movements. He co-edited the book *Climate Change, Carbon Trading and Civil Society* (2007), edited *Durban's Climate Gamble* (2011), and authored *Politics of Climate Justice* (2012).

Eugenia Cossa is Professor and Director of the Faculty of Education at Eduardo Mondlane University in Maputo, Mozambique.

Rehana Dada is a Masters student at the Centre for Civil Society, University of KwaZulu-Natal, Durban, South Africa, as well as a journalist and activist. She co-edited the book *Climate Change, Carbon Trading and Civil Society* (2007).

Patricia Figueiredo Walker is Project Coordinator of the IDRC/DFID-funded project entitled "Strengthening the role of civil society in water sector governance towards climate change adaptation in African cities – Durban, Maputo, Nairobi" based in the Faculty of Environmental Studies, York University, Toronto, Canada.

Mary Galvin is Executive Director of the NGO Umphilo Wa Manzi (Water is Life), based in Durban, South Africa. Umphilo's mission is "To engage in advocacy and action research around issues related to water and sanitation, the environment, and health in South Africa."

Wahu Kaara is Executive Director of the Kenya Debt Relief Network (KENDREN), based in Nairobi, Kenya. KENDREN's motto is "Towards a Free, Just and Equitable Society," and its mission is "To become the national platform for collective action committed to the total cancellation of Kenya's public debt and its effective management and to empower the public on issues of economic justice and sustainable development through debate, advocacy and organizing."

Erika Mendes and **Tiago Esmael** are activists with the NGO Justiça Ambiental! (Environmental Justice!), based in Maputo, Mozambique. JA!'s mission is "To engender a culture of civil action in Mozambique both through actions to protect the environment, and by actively engaging in developmental decisions pertaining to issues of environmental justice, here in Mozambique and throughout the rest of the world."

Saquina Mucavele is Executive Director of the NGO Mulher, Gênero e Desenvolvimento (MuGeDe: Women, Gender, and Development), based in Maputo, Mozambique. MuGeDe's mission is "To promote a healthy environment through the combined efforts of different social groups, including the government. To focus on the environment and rural development from a gender perspective."

Lushendrie Naidu is the former Projects Officer with the South Durban Community Environmental Alliance, an NGO based in Durban, South Africa. Its mission is "To protect and ensure an environment conducive to positive well-being of the citizenry; to promote 'clean' development that aids local job creation; to protect resources for present and future generations and to ensure development that is not detrimental to the health of residents, especially children."

Simphiwe Nojiyeza is a PhD student at the Centre for Civil Society, University of KwaZulu-Natal, Durban, South Africa, and an activist with the organization WET Africa (Waterways Environmental Transformation).

Romanus Otieno Opiyo is a Lecturer in the Department of Urban and Regional Planning at the University of Nairobi, Kenya. His research focuses on planning and climate change.

Patricia E. Perkins is Professor in the Faculty of Environmental Studies, York University, Toronto, Canada and directed the project on climate change and water management in Durban, Maputo, and Nairobi discussed in this book. Her research focuses on climate justice, participatory environmental governance, community development, and feminist ecological economics.

1 Introduction

The importance of community-based organizations for equitable water governance in times of climate change

Patricia E. Perkins

Climate change affects all of us, wherever on Earth we live. Weather is changing and storms are increasing in intensity, stressing agriculture and infrastructure. As carbon dioxide and other gases emitted by burning fossil fuels cause the earth to heat up, more water evaporates into the atmosphere, and the sun's heat, caught in the thickening atmosphere, drives powerful winds and weather cells, causing droughts, storms and floods more extreme than humans have ever witnessed before.

International strategies for addressing climate change are in disarray. Even if a global consensus had been reached on the United Nations Kyoto Protocol, and its provisions had been implemented and had succeeded in reducing global carbon emissions to 1990 levels, global temperatures would still have risen at least 2 °C, meaning that much of Bangladesh, Florida, Manhattan, and many small island nations, could be underwater by 2050. But the Kyoto Protocol was *not* agreed upon; the wealthy nations which have benefited from fossil fuel based economic growth over the past 200 years refuse to reduce their consumption, and the surging growth in the rest of the world also requires fossil fuels to maintain its impetus. It now appears certain that global temperatures will rise by at least 3 °C. The complicated financial and carbon-trading mechanisms promoted by the United Nations Framework Convention on Climate Change, the World Bank, and other global institutions are far too bureaucratic, weak, internally-inconsistent, and far-fetched to represent meaningful solutions to climate change. And the housing, health, and livelihoods of marginalized people worldwide are already being threatened by climate-change-related impacts.

This means that the marginalized in every community have both the necessity and the means to take direct action. Their lived experience gives them expertise in how priorities should be set to address climate change. Their knowledge and views must be part of local, regional, national, and international governance—including urban planning and housing, water management, infrastructure, agriculture, health, and finance policies. Women's special knowledge, in particular, is crucial, because of women's socially-constructed responsibilities throughout the world for fetching water, cooking, cleaning, agriculture, healthcare, and maintaining housing—all water-dependent activities. Given the immediacy of humanity's common climate problem, and the shortage of resources to address it, we must address these challenges efficiently and develop resilience to face them—at the community level.

What are some of the challenges that low and middle-income countries face as a result of climate change?

> (M)ost of the world's urban population live in cities or smaller urban centres ill-equipped for adaptation—with weak and ineffective local governments and with very inadequate provision for the infrastructure and services needed to reduce climate-change-related risks and vulnerabilities. A key part of adaptation concerns infrastructure and buildings—but much of the urban population in Africa, Asia and Latin America have no infrastructure to adapt—no all-weather roads, piped water supplies or drains—and live in poor-quality housing in floodplains or on slopes at risk of landslides. Most international agencies have long refused to support urban programmes, especially those that address these problems.
>
> (Satterthwaite *et al.* 2007: vi)

The World Wide Fund for Nature's (WWF) review of the scientific literature on climate change impacts in East Africa underscores the importance of changes in precipitation and weather, sea-level rise, water availability, and other water-related problems, and states that, "Climate change impacts have the potential to undermine and even undo progress made in improving the socio-economic well-being of East Africans. . . . For every USD$1 spent preparing for disaster, USD$7 is spent recovering from disaster" (WWF 2006: 1 and 9). The equity implications of climate change impacts are underscored in a third recent report which addresses urban flooding and the rights of the urban poor in Africa:

> Climate change will increase the vulnerability of the urban poor throughout Africa. . . . The management of localized flooding due to inadequate drainage should be undertaken by local communities themselves. . . . Local voluntary groups, assisted by national or international NGOs [non-governmental organizations] and with support from both local government and national disaster reduction organisations, could be highly effective. . . . Local authorities are (also) best placed to cope with flooding from small streams whose catchment areas lie almost entirely within the built-up area. . . . Poor people's participation is a must.
>
> (ActionAid 2006: 2 and 6)

In the coming years, countries around the world will face increasingly severe problems stemming from global climate change. While the details vary from place to place, the impacts are especially grave for marginalized people, whose access to food, drinking water, and safe shelter is most often threatened due to fluctuations in rainfall and temperature, and to extreme weather events. Climate change thus worsens problems related to poverty and inequality, which persist everywhere despite decades of "sustainable development" programming.

Climate change adaptation cannot be standardized. The details of each particular community's situation—ecological, social, political—affect its own priorities. How does each community organize socially and politically to meet the material and weather-related changes that affect people's livelihoods? How are the needs of the most vulnerable addressed in local communities? The chapters in this book describe how civil society organizations (CSOs) which are already working in low-income neighbourhoods can incorporate climate change education and adaptation into their activities. University students, by working as interns with the CSOs, can learn how to do community outreach and share their academic knowledge, while helping the CSOs to document their organizing initiatives and workshops. When local CSOs have connections within universities, this strengthens them politically and builds communication channels between vulnerable people—the experts on climate change impacts—and government officials who are responsible for developing climate adaptation policies.

The goal of this book is to contribute to a "bottom-up" response to climate change by describing and demonstrating how community-based climate change strategies can be initiated, and how international collaboration can strengthen and foster such grassroots initiatives. Thus, this is an action-oriented book about the *process* of climate change response and adaptation, starting in local communities.

Climate change and water governance in Durban, Maputo, and Nairobi

According to the Intergovernmental Panel on Climate Change (IPCC), "Africa is one of the most vulnerable continents to climate change and climate variability. This vulnerability is exacerbated by existing developmental challenges such as endemic poverty, limited access to capital, ecosystem degradation, and complex disasters and conflicts" (IPCC 2007). Income inequality in South Africa, Mozambique, and Kenya is among the largest in the world; in all three countries, equity struggles related to water are growing in social, political, and ecological significance, which is both a symptom and a cause of urban vulnerabilities related to climate change.

In Maputo, Mozambique, climate change is causing coastal erosion and periodic flooding along scenic coastal roadways; saltwater intrusion, wind erosion, and desertification in urban food-producing areas; flooding in coastal slum areas; degradation of water quality in wells and potable water scarcity; and the destruction of mangroves and threats to the locally-important shrimp fishery. There are clear signs that the sea level is rising, which has led to the need for expensive coastal management efforts in Maputo municipality. On three offshore islands mangroves are disappearing, water quality is declining, and desertification and erosion are increasing (UN Habitat 2010: 2). The United Nations Habitat Cities and Climate Change Initiative, which has begun a pilot project in Maputo, emphasizes local government capacity-building, policy dialogue, climate change awareness, public education, and developing coordination mechanisms between all levels of government as priorities to help address these risks. Mozambique's national water law (1991) considers all water as state-owned, to be governed by the state for the benefit of the population, with water access for people, sustainability, and stakeholder participation as priorities. Four water basin committees have been established in Mozambique, with seats for civil society representatives.

As in Mozambique, South Africa is implementing watershed committees or "catchment management agencies" (CMAs) to decentralize decision-making and create a framework for integrating the needs of all stakeholders in water governance. Durban's municipal government has already developed a local climate change adaptation strategy; like Maputo, Durban faces coastal floods and storm surges related to sea-level rise, hotter temperatures and heat waves, changed rainfall and storm patterns, slum flooding and reduced drinking water supplies due to climate change. Environmental education and confidence-building are recognized as crucial needs; for example, the Inkomati CMA has initiated outreach programmes targeting rural poor, emerging farmers, women, and youth. Grounded participatory research leading to accessible public education and responsive community-based programmes with CSOs are needed to help address these significant water governance challenges. This type of action research is well developed in Durban, partly due to the work of the Centre for Civil Society at the University of KwaZulu-Natal and its partner CSOs.

In Nairobi, severe infrastructure needs are being exacerbated by water supply fluctuations and slum flooding related to climate change. Just as in Maputo and Durban, environmental awareness and education leading to more equitable governance processes are needed. As noted by the Kenyan delegation to the 2006 UN conference on climate change in Nairobi, Kenya's adaptation focuses include education, good governance, human resources development and training, institutional capacity building and management change, public finance improvement, and better national resources management. Nairobi, one of the largest and most complex cities in the world, provides a challenging arena for participatory governance research.

Democratic mediation of equity conflicts related to water, and sustainable long-term management of water resources in the face of climate change, requires public participation, in particular by low-income marginalized women—the experts. But many people's lack of awareness of how water governance institutions function, and inability to participate for a range of reasons, mean that low-income people are nearly always underrepresented.

"Strengthening the role of civil society in water sector governance towards climate change adaptation in African cities—Durban, Maputo, Nairobi" is a three-year project linking African CSO and university-based partners in these three cities. Its goal is to improve watershed governance for climate change adaptation and enhance resilience and adaptive capacity of vulnerable and marginalized groups, especially women, by developing methods for strengthening the voices of marginalized people and especially women in water governance and linking communities with other water actors to help bridge the knowledge gap in local water management and enhance a broader social perspective on water, climate change, and sustainability.

This project, which started in 2010, is supported by the Climate Change Adaptation in Africa (CCAA) programme—a joint initiative of Canada's International Development Research Centre (IDRC) and the United Kingdom's Department for International Development (DFID). This project's methodology includes collaboration between students, NGOs, and academics as well as community-based research and environmental education. Project partners based in universities and several NGOs in Nairobi, Maputo, and Durban are working together to achieve the following objectives:

- To characterize the institutional framework for urban water governance in the three cities, and explain how different actors within this framework cope with climate change and variability.
- To identify and test viable alternatives for enhancing civil society's role towards adaptation to climate change and variability by vulnerable groups (e.g. by developing education, training, and awareness programmes).
- To share widely the knowledge generated for potential adoption by other cities in Africa.

The project is being implemented by the following community-based NGOs in Africa: Kilimanjaro Initiative (KI) and Kenya Debt Relief Network (KENDREN) in Nairobi; Women, Gender and Development (MuGeDe) and Justiça Ambiental (JA) in Maputo; and Umphilo waManzi (Water for Life) and the South Durban Community Environmental Alliance (SDCEA) in Durban. The University of Nairobi (Nairobi), Eduardo Mondlane University (Maputo), the

Centre for Civil Society at the University of KwaZulu-Natal (Durban), and the Faculty of Environmental Studies at York University in Toronto, Canada provide academic research coordination and student supervision for this project. Their chapters in this book describe the starting-points for all of these partner organizations in their approaches to climate change-related work.

Participatory climate change awareness-building

In general, all the partners share a commitment to focusing on low-income areas of each city; these tend to be most severely affected by periodic flooding and other climate change impacts. Residents of low-income areas often lack the ability to protect themselves against the impacts of extreme weather events. The project includes training and research sponsorship for students and faculty in the partner universities; support for community-based research, workshops in low-income communities and secondary schools, curriculum and materials development, and skills development within the partner NGOs; training of environmental educators and organizers; contributions to the pool of experienced and qualified community workers in each country; strengthening of all the partner institutions' capabilities to carry out international projects; and contributions to the international literature and professional knowledge concerning water issues, environmental education techniques, and community organizing for improved civil society involvement in governance. The networks being built extend from local and community-based linkages through regional and national-level policy groupings to international academic and policy networks on civil society, watershed management, and governance.

The political process of climate change policy development and implementation depends on the interchange between civil society groups, researchers generating information on current realities, and government. We are attempting to challenge the conventional notion that only educational institutions "produce" knowledge. Understanding community needs and what helps particular civil society groups to see and act to strengthen their role in democratic governance, for example, is something in which community organizations and CSOs have eminent expertise. This collaborative approach, also known as participatory action research (PAR), is broadly defined as "research by, with, and for people affected by a particular problem, which takes place in collaboration with academic researchers. It seeks to democratize knowledge production and foster opportunities for empowerment by those involved" (Kindon *et al.* 2008: 90).

Partnerships between academics and non-academics can be especially productive and effective. This type of partnership encourages and allows the partner CSOs to reflect on and analyze their activities and to document "learning" more systematically than they are often able to do, by bringing student researchers into the CSOs as collaborators/interns. It also encourages universities to be more pragmatic about teaching and research, and to "field-test" approaches towards community organization, equity, and popular education. Students committed to the project's goals of supporting participatory engagement by local people in municipal water decision-making are given practical opportunities to develop their skills, as a way of hastening each city's climate change preparedness.

The integration and meaningful participation of women in formal decision-making processes is especially important in times of climate change, given women's gendered responsibilities for household food, fuel, and water provision as well as healthcare, childcare, elder care, and community supports. This requires attention to women's adaptive capacity and special supports for women's resilience and ability to cope with climate change.

Specific examples of how climate change responses combine well with gender-aware community organizing, all of which are now underway, include the following:

- The KI, a youth-focused NGO, has upgraded a sports field in Nairobi's Kibera slum, on the banks of the Nairobi River, which will prevent housing from being flooded during extreme weather events. In addition, KI organizes community forums on sustainable water management and environmental education, as well as community and river clean-ups. Young women's leadership is central to their organizing. (See chapter 4.)
- In Durban, women activists from Umphilo waManzi and the SDCEA coordinate "learning journeys" where government officials visit low-income neighbourhoods to hear about local women's experiences with flooding, sanitation, and other types of climate change stresses, which helps them to bring these views into policy discourse. (See chapters 5 and 10.)

- Maputo university environmental education students work with intermediate school youth on after-school activities related to climate change. Most participants are women. (See chapters 6 and 9.)

Community-based education and organizing are fundamental to creating the conditions for local knowledge to be shared and utilized, through equitable democratic participation. Building inclusive governance structures and strengthening the role of civil society, especially women, in water governance are essential components for addressing climate change vulnerability and fostering resilience and sustainability in urban centres as well as rural areas. According to the Intergovernmental Panel on Climate Change, "adaptation is shown to be successful and sustainable when linked to effective governance systems, civil and political rights and literacy" (Parry *et al.* 2007: 151). Non-governmental CSOs in the Global South have expertise in such initiatives, which is potentially transferable to other places—including places in the Global North.

Community-based environmental education initiatives which are relevant and interesting for local residents and increase their job opportunities, knowledge of watershed issues, understanding of basic political and ecological principles, and their confidence to express and act on their views, can serve as the basis of a climate change intervention approach which is progressive, constructive, and democratic. This, in turn, increases the resilience and sustainability of watershed and climate change decision-making processes. It also lays the groundwork for community organizing and extension of the environmental education activities to larger constituencies in local areas affected by climate change. Such grassroots initiatives—and the global sharing of ideas on how to design and implement them, freely available for adaptation in other places—stand in contrast to top-down climate change adaptation mechanisms controlled from the Global North. In this sense, climate justice is a new manifestation of the bottom-up perspective in Development Studies more generally. Furthermore, it is a movement which "best fuses a variety of progressive political-economic and political-ecological currents to combat climate change" (Bond and Dorsey 2010: 286).

Climate justice—addressing the impacts of climate change on the poorest first—is a powerful imperative at every level, from the local to the global. Civil society groups worldwide are using online and in-person networking tools to share ideas on how to promote climate justice, to obtain funding, and to press politically for policies addressing the needs of marginalized people. This bottom-up movement builds resilience in the face of the social and political repercussions of extreme weather events.

Our strategy in designing and carrying out this work has been to start where groups and people are, jointly analyze the local situations, name and build on partners' strengths, collaborate in creating new initiatives, share our results at each stage, build bridges between academic and community groups through student work and exchange visits, and share perspectives to foster a global vision, gaining energy and ideas from each other.

Our work emphasizes that progressive participatory governance is needed in order to achieve sustainable water resources management and community resilience in the face of climate change. This type of progressive governance requires the recognition that water resources are a "commons" rather than a "commodity".

In the following chapters, academics and activists discuss climate change and water challenges in Durban, Maputo, and Nairobi and how their own current work is helping to address these challenges. Each chapter describes the current work of each partner organization related to water and climate change, including a description of the apparent local challenges and how they are being addressed. In the concluding chapter, we summarize our joint strategy and describe our frameworks for supporting each other as we move forward.

Our goal in presenting these stories is to show the diversity and creativity of grassroots, non-governmental responses to climate change in urban Africa today. Our hope is that these stories will resonate with and inspire others who are working together to build frameworks for community resilience and action in the face of humanity's greatest challenge to date: climate change.

2 Economic, ecological and climate justice—literacy and politics, from rights to commons

Wahu Kaara

As an actor in the trenches of the Global Justice Movement and specifically in our work around debt, aid, and economic governance, KENDREN—the Kenya Debt Relief Network—continues to hold a strong position against International Financial Institutions (namely the World Bank, the International Monetary Fund and their generic versions, the International Finance Corporation, and the African Development Bank). We work in a policy space that spans African economies and development models.

KENDREN was founded at the time of the 1999 Jubilee movement for cancelling international debt, or dropping the debt. We participated in the World Summit for Sustainable Development in 2002, and helped to create a national consultative process. KENDREN holds debt clinics in local communities to discuss issues with the poor. People may be poor, but they still owe debt—an average of 30,000 shillings is the amount of debt service that Kenyans pay to international creditors every month. Most people don't even know they have this money! We introduce the subject in a participatory way, and promote economic and ecological and climate justice literacy.

We interact with those who are most vested in these issues: the urban poor, because they hold the power for social transformation. They are very integrated in the money economy; they cannot eat without money and they need to pay rent, but they do not have guaranteed sources of income. They have to be innovative to secure income. So there is a lot of energy in poor communities. KENDREN's greatest objective is to build networking connections with those who are agents of social transformation, and to build social movements or join others in strengthening the social movements that every community does have, since each community has its own issues and problems on which they need to act. Communities must be the driving force and power behind their own interests.

We noticed that when we started talking with participants about ecological debt, people were confused. But the economic debt has a very clear relationship with ecological debt, which is why KENDREN is involved in climate change and environmental issues. We participated in the Climaforum at the COP 15 climate conference in Copenhagen, drawing linkages between economic justice, ecological justice, and climate justice.

Clima and Social Forums mobilized the communities to have village assemblies on debt, and also to discuss debt and climate justice in their own congresses (their constitutional assemblies). The issues discussed in these assemblies are related to the environment and climate change; the economic literacy of the people is beginning to manifest itself into literacy on environment and climate justice, as well.

The international "debt relief" movement is motivated by the conviction that so-called "debtor" nations have in fact funded economic growth in the so-called "creditor" nations for centuries, first via colonialism and the slave trade, and subsequently via huge interest and foreign exchange payments, and resource exports, to banks, governments, and transnational corporations, which dwarf the aid and loan transfers going the other way. This analysis is easily extended to climate change and the "climate debt": those who have consumed the vast majority of climate-change-causing fossil fuels are in fact those who owe a "climate debt" to the rest of the world's population, who are suffering from the effects of climate change. The way to bring about "climate debt relief," just like "debt relief," is to share the burdens and the benefits more fairly across time and space.

Our logic in analyzing these issues is informed by the fact that, inherently, there exist particular "commons" that are shared among all humans and other beings. However, different people have different access to these commons. The economic logic that contextualizes societal relations and dynamics from the lenses of a neo-liberal market model is a destructive construct inasmuch as it pursues a profit motive at the expense of human progress and societal advancement. This relates to financial debt, climate debt, and also to water governance. It is critical to move the discussion beyond the "rights" rhetoric to a more persuasive discussion on "commons."

Let me illustrate. Nairobi is a city of slightly more than four million people. By all indicators, it's a robust city in more ways than one, but it is also a city that is caught up in the contradictions of inequality. With regard to Nairobi's water governance, today we boast of the leading company on water management, The Nairobi City Water and Sewerage Company—a water service provider and subsidiary of the Nairobi City Council. This company is a spinoff from when water was managed by the Nairobi City Council. And when I say management, I am referring to the sourcing, distribution, maintenance, and management of water. Core arguments revolve around the inefficiency of Nairobi City Council and the partial privatization of the water company after 2002, under the tutelage of the World Bank and other donors, which assigned the management and provision of water and sewerage services to Water Service Providers (WSPs).

Incidentally, the main thrust informing this decision was that state enterprises would fail when involved in complex business like water. As a result the best framework is to have the state as the regulatory framework, leaving the day-to-day management and distribution issues to the private company.

This is indeed a compelling argument when it's analyzed or presented from the position of those who can afford to pay for it.

The "rights" to "commons" argument I make is hinged on the fact that "water is life." It may not be a "right" per se in the context of paying for it, but in terms of a need for all (across class, purchasing power and address) there inherently lies the "commons" will to access both clean and consumable water for all.

The argument that the state cannot perform this function fails to contextualize the role of the state as not only a distributor and regulator of competing needs in society, but most importantly as a defender of social equity. Or perhaps, this argument represents an example of "mindset capture" by the forces of the market.

Living examples: scenario 1

Huruma is a densely populated region of Nairobi, located approximately ten kilometers east of the central business district, and is home to millions of low-income working-class families, casual workers, and others in the majority who are eking out a city life. Arguably, it is not a slum according to the typical definition, though the district includes various dense low-income neighborhoods—Mathare, Huruma Flats, Korogocho, and Kariobangi.

Key contestations here revolve around "illegal" water connections and a vibrant sub-economy that is characterized by mushrooming "car washes," water vending, communal showers and toilets. In this regard, there exists a buoyant "underground" economy that rests on the ingenuity of local residents to tap into the city water grid to run local enterprises. The sanctity of doing so is still debatable, but it has to be recognized that this form of "local economy" goes a long way in securing livelihoods for the majority of the urban poor in these regions.

It's equally a powerful testament to local innovation in the wake of the debilitating effects of being at the base of the societal pyramid. The argument here is that the water should equally serve to help residents secure their livelihoods, in just the same way that large enterprises use tons of water for their production. Thus, the battles in Huruma are less about having enough water per household, and more about the economic value that can be derived from its utilization. Although the battles have all too often been bloody, a certain sense of an uneasy agreement between the residents and the water company seems to exist. For a certain level of access, the residents will protect the water pipelines from vandalism. Thus, an informal unity of purpose exists; the company needs to keep its grid going and the residents can continue with their water-based businesses.

Living examples: scenario 2

Kibera, on the other hand, holds a development record as the largest slum in sub-Saharan Africa. It is the site of a large influx of development work—both international and local organizations have made inroads here, in sectors ranging from health to water to the World Bank's own slum upgrading project.

Unlike Huruma, Kibera seems to be in tune with its plight. Despite the large concentration of people and the hazards related to this, Kibera's access to clean and safe drinking water is laudable. Kibera is in a sense a poster child for development, since most residents do have access to clean and safe drinking water.

But here is the catch: in most instances, the water is piped to common points from which the residents draw the water, ideally managed by a local community group. A 20-liter can retails for about 5–10 Kenyan shillings, which in effect means that water consumers in Kibera on average spend more on water than their counterparts in relatively well-off neighborhoods.

In both these examples, slum dwellers have less access to water than richer people. But water, which falls from heaven, is a necessity of life, and so the "commons" framework is highly appropriate for considering water access—much more so than a market framework.

Through KENDREN's economic literacy programs, we began working years ago with people in Huruma. We have been energized by their commitment to addressing environmental issues, such as garbage collection. Through the economic literacy programs, the people begin to feel that they need to take control of their lives and take some responsibility if anything is ever going to change in their communities. It is an amazing process that has brought even those who have experienced crime to develop a way of alternative action and alternative thinking. They began to organize houses to pay the neighborhood's young people to collect and dispose of garbage. This created a big question between them and the city council, since they needed to interrogate the by-laws, and sought legal advice on behalf of the community. In fact, this process has linked all of Nairobi's informal settlements or poor settlements, because garbage is a problem everywhere.

Critical also is that Kibera and Huruma are part of the Nairobi River System, which is very polluted. In recent years, the cleaning of Nairobi River has become very important. The United Nations Environment Programme, based in Nairobi, has a proposal for cleaning up the river and making it habitable for local residents and families. It is full of brilliant ideas as to how to transform Nairobi—how to promote tourism, sporting events, concerts along the river, street soccer, etc. These activities can make this community a manifestation of the future of Kenya.

Building social movements becomes very critical, along with linkages of knowledge and ideas that lie within the communities. This knowledge is a living library that is never exhausted; people are not difficult to follow because you meet them everyday and hear their daily stories and testimonies. We must learn to make linkages with formalized knowledge from the universities, and formal research, and begin to document the knowledge of communities.

Climate change is giving poor communities an opportunity to go into depth and participate with the politics of economics, the politics of environment, and the politics of climate change itself—especially within the international discourse framework of mitigation and adaptation.

Conclusion

Faced with the gigantic growth of African cities, a better-defined and broadly-accepted strategy on water governance needs to be sought. This must take into context the varied needs of the different population segments which cities like Nairobi present. Otherwise water will be a big challenge and an open site for new contestations into the future, and these contestations will be heightened by climate change.

Second, the role of the state in water governance must be rethought. After structurally adjusting our economies for over two decades, the verdict is in: the state still has a great role to play, especially by setting the regulatory framework and enforcing notions of equity and equality.

Third, the fundamentals of the new discourse on water must move beyond the rhetoric of "rights" to "commons," particularly because fundamental freedoms are so important that all people must know that society's structures involve them as actors and collaborators in decision-making.

Most importantly, the paradigm must shift from blind market orientation to promotion and recognition of the key role of the state in adjudicating societal concerns. And by this, I do not mean the cosmetic application of liberal democracy, but a more in-depth indigenous approach to local problem-solving. In this regard, cross-learning across nations will be vital in defending the shared common wealth and the "commons" of water as the foundation of our common subsistence.

Kenya is so fortunate because our new constitution is already in place. The constitution establishes the principle of subsidiarity and brings the process of decision making to where people are. This will help to locate the reality that there is alternative thinking, alternative action, and alternative ways of moving forward—from the very dispossessed, from the ashes. Community knowledge should be respected in matters of mitigation and adaptation to climate change. We cannot deny that people can lift themselves out of their dispossession.

3 Climate, gender and class justice in Africa

Patrick Bond

'Climate justice' fuses progressive political-economic and political-ecological currents to combat the most serious threat humanity and most other species face in the twenty-first century. Climate justice arrived on the international scene as a coherent political approach in the wake of the failure of a more collaborative strategy between major environmental non-governmental organizations (NGOs) and the global capitalist managerial class. But it is important to ensure that the gendered and class distinctions in addressing climate politics be foregrounded, the subject of this essay.

Climate justice activists make demands that simply will not be met at the Conference of the Parties (COPs) to the United Nations Framework Convention on Climate Change (UNFCCC): a 50 per cent greenhouse gas emissions cut by 2020 and 90 per cent commitment for 2050; payment of a rapidly rising 'climate debt' (in 2010, damages to Pakistan alone amounted to US$50 billion); the decommissioning of the carbon markets so favoured by elites; and massive investments in renewable energy, public transport and other transformative infrastructure. As a result, climate justice struggles are necessarily retreating from a naively overambitious global reform agenda (politely asking UNFCCC delegates to save the planet) and instead promoting direct action, as witnessed in sites across the world:

- Niger Delta women, Environmental Rights Action, and MEND (Movement for the Emancipation of the Niger Delta) halted the majority of oil exploitation in the Delta at peak in 2008.
- Canadian anti-tar sands campaigning unites green and indigenous activists, and Quebecois fought successfully against shale-gas.

- British Climate Camp and 'Crude Awakening' blocked the Coryton oil refinery in 2010 – and even an MI5 spy couldn't break that network.
- Australian Rising Tide and allies regularly block Newcastle coal exports.
- Norwegian environmentalists and Attac won against the state oil company in the Lofoten region in 2011.
- In Amazonian Ecuador, indigenous activists and Acción Ecológica have so far halted oil drilling in Yasuni National Park.
- 'King Coal' is threatened in the US: mountain top removal nearly halted in Appalachia; the Navajo Nation forced cancellation of a Black Meza (Arizona) mine permit against the world's largest coal company, Peabody; Powder River Basin (Montana, Wyoming) farmers and ranchers fight coal expansion.
- Derailing US coal energy: nearly all 151 proposed new coal power plants in the Bush Energy Plan have been cancelled, abandoned or stalled since 2007; the key community forces are the Indigenous Environmental Network, Energy Justice Network and Western Mining Action Network, plus the Sierra legal defense team.
- Preventing waste incinerators in the US: there is strong opposition to new waste incinerators (more carbon-intensive than coal and a leading source of cancer-causing dioxins).
- Global Alliance for Incinerator Alternatives won a major Detroit victory and helped unite a world wastepickers movement with climate narratives.
- Chevron's expansion in Richmond, California was defeated.
- Mega Hydro was undammed at Klamath River as indigenous communities defeated Pacificorp Power.
- Resilient communities are being built through local action, including frontline communities winning campaigns linking climate justice to basic survival (e.g. the Oakland Climate Action Coalition promoting Just Transition).

Fused as climate justice, these inter-related and often overlapping (although sometimes conflicting) traditions are mainly aimed at building a mass-based popular movement bringing together 'green' and 'red' (or in the U.S., 'blue-green') politics. This entails articulating not only the urgency of reducing greenhouse gas emissions but also the need to transform our inherited systems of materials extraction, transport and distribution, energy-generation, production of goods and services, consumption, disposal and financing.

The climate justice organizations and networks offer great potential to fuse issue-specific progressive environmental and social activists, many of which have strong roots in oppressed communities, in bottom-up linkages of organic climate activism. But climate justice movements across the world have not solidified a coherent set of tactics, much less strategy, principles, ideology and foundational philosophy. Climate justice is by no means a global 'movement'.

However, there have been early efforts to raise the profile of climate justice beyond mere attendance and protest at the UNFCCC COPs. At a landmark conference in Cochabamba, Bolivia from April 19–22, 2010 (Earth Day), more than 30,000 Bolivians led by host Evo Morales and grassroots indigenous, community, feminist, and environmental movements were joined by genuinely solidary environmental, social, labour and NGO forces. This meeting set in motion a much more serious transnational climate justice approach, based not upon the illusion that the UN will address the climate crisis anytime soon, but instead upon more serious, pragmatic strategies. These can come only from a much richer merging of social and ecological rights discourses (for what such narratives are worth), and choices of unifying targets (such as fossil fuel companies, carbon traders, and the World Bank).

Everywhere, climate justice organizers can find excellent local climate change targets to raise consciousness and effect emissions cuts, with direct action against major greenhouse gas sources or large-scale corporate fossil fuel consumers. Two particularly good sites for climate activists are, first, the public utility commissions which control pricing and electricity generation techniques (and hence coal-fired and nuclear power plants); and second, the municipal or regional planning commissions which give the go-ahead to suburban sprawl and all manner of other climate-threatening projects.

Diversity and climate justice politics

No matter that the Climate Justice Now! component movements are disparate; so too are the forces that moved from sophisticated critique of carbon trading in South Africa to a broad-based campaign against the World Bank's largest project loan, shaking the energy establishment in the process. These are the kinds of organizations and arguments that link spatio-temporal resistances amongst diverse eco-social forces during a period of austerity, civil society weakness, and repression. The agents of social and environmental change can take advantage of neoliberalism's still-discredited ideological status and demand from the next global and national negotiations a strategy for justice, not one based upon commodifying carbon.

But to do so the climate justice activists still need to generalize an innovative critique, one that emerged over time as the global emissions-trading strategy rose from birth in 1997, peaked in 2008, and attempts a last-gasp resurrection after the economic crisis and climate negotiations breakdown. The frenzied failure of elite climate politics should make bottom-up alternatives much easier to advocate, including globally-coordinated actions against destructive projects (such as international credits for South African coal).

From the wider, deeper, and increasingly common critique of the Kyoto–Copenhagen–Cancun–Durban cul-de-sac, will come more confidence in the types of analyses, strategies, tactics and alliances associated with climate justice politics. The challenge is to aggregate experiences from the grassroots so as to move into alignment with the Cochabamba conference, and in turn, to generate a formidable red-green force uniting radical governments and the popular movements that will keep them accountable within global-left initiatives. In South Africa, the experiences of national interactions in global climate negotiations reinforce a sense of how the politics of scale and space can be distorted, so that policies patently against the interests of a country's mass-popular constituencies are adopted. But likewise, South African climate justice politics from below suggest ways forward that, while not yet sufficiently strong to declare victory, really do offer the only hope for the way forward.

Ecofeminism and climate justice

What Africa and the world must yet grapple with is the fusion of ecofeminist and eco-socialist ideological imperatives, within a revitalized climate justice politics. It is enlightening to contextualize ecofeminist insights into climate politics within a concrete case study in which class, race and environmental oppressions are inextricably intertwined with gender. During the 1960s to 1970s, it was recognized amongst radical South Africans that the apartheid system's discrimination against black people was rooted in the corporations' need for migrant labour, fusing race-class oppression.

Behind the typical black man who laboured in the mines throughout the first century of organized mining and plantation agriculture, prior to Nelson Mandela's election in 1994, was a woman. She provided three hidden and un-costed subsidies within the 'care economy,' responsible for labour power's cheap reproduction. First, in rural Bantustans – the ecologically-degraded apartheid 'homelands' – women raised the migrant worker through childhood, as the state was non-existent or merely a religious mission station. Household reproduction was never subsidized, unlike urban residents who in many cases had access to state childcare and school systems. Second, rural women were compelled to look after sick workers who were tossed back home until they recovered, due to the lack of health insurance, as offered by states and companies in the West after workers battled long and hard. Finally, when the male worker was too old to work and returned to the Bantustans without adequate pension support, the women again took on the responsibility for care giving. These are the women most likely to suffer additionally as climate change hits southern Africa.

Of course, it's not just a matter of apartheid capitalism. The reproduction of global labour power has been universally subsidized by women's unpaid work. Neoliberal policies and corporate power have resulted in labour outsourcing, casualization and informalization. In times of recession, global capitalism becomes much more like apartheid: predatory against women and the environment. With life more precarious as a result, women are the safety net for household reproduction, in addition to being the most vulnerable and disposable of all labour sectors. But they have also been the driving force in resisting this process, overcoming micropatriarchy within communities and leading most of South Africa's grassroots campaigns on issues such as water decommodification, access to AIDS medicines and other successful strategies to enlarge or defend the commons and sustain life.

Drawing on evidence from southern Africa, Rosa Luxemburg demonstrated this tendency in her own analysis of imperialism back in 1913: 'Accumulation of capital periodically bursts out in crises and spurs capital on to a continual extension of the market. Capital cannot accumulate without the aid of non-capitalist organizations.'

Increasingly, such non-capitalist life arrangements rely upon women and the communities that they guide. And yet on the other hand, Luxemburg continued, capitalism cannot 'tolerate their continued existence side by side with itself. Only the continuous and progressive disintegration of non-capitalist organizations makes accumulation of capital possible' (Luxemburg, 1968).

Luxemburg would not have been surprised at how the destructive force of capital drives the migration of men, spreads HIV/AIDS and causes rising domestic violence. Such disintegration is always contested by women's personal strengths and mutual aid systems as well as other anti-/non-capitalist reactions, plus campaigns – successful in South Africa (unique on the continent) – to guarantee reproductive health-care, including the right to a safe abortion. But increasingly, with climate-related migration, the challenge is more formidable because the progressive disintegration of the natural world by capitalist expansion and emissions is what makes the next round of accumulation possible, in the short term, but entirely dubious over the medium and long term.

Teresa Brennan made the link from the household scale to climate change, the biggest crisis women will face in the coming decades. She argued that, like the need to end Bantustan migrant labour systems, rearranging spatial and re/production arrangements is crucial to ending the unfair role of women in subsidizing capitalism's destructive irrationality. In her 2003 book, *Globalization and its Terrors: Daily Life in the West*, Brennan wrote: 'The closer to home one's energy and raw material sources are, the more one's reproduction costs stay in line: paid and domestic labour will be less exploited, the environment less depleted' (Brennan, 2003). The need now to limit the 'distance over which natural resources can be obtained' is obvious given how shipping, trucking and air transport contribute to carbon emissions.

This is one of the insights an ecofeminist political economy gives climate strategists such as Nicola Bullard of Focus on the Global South. A typical debate with neoliberals is over whether globalization of industry has helped break up feudal-patriarchal relations, drawing women out of oppression into Mexican maquiladoras or Bangkok sweatshops. Such export-led growth is now an increasingly untenable 'development' strategy, and in any case always generated extremely uneven development, drawing on the women's care economy for its hidden subsidies.

Bullard likens the climate negotiations to those of the World Trade Organization: 'By and large, countries are defending their narrow economic interests and the rich countries in particular are trying to grab the last slice of the atmospheric pie' (Bullard, 2009). Although the Kyoto Protocol is deeply flawed, especially the low targets and reliance on market mechanisms, Bullard asserted that attempts by the US to get rid of Kyoto are dangerous. 'It is critical to retain the rich countries' legally binding commitment in any future agreement and any alternative that could emerge at this stage would be much worse.'

Bullard breaks down the climate policy narrative into three discourses: business as usual, catastrophism, and climate justice. The first comes from business and most northern governments while the second is advanced by some smaller and vulnerable countries as well as many NGOs. Catastrophism also 'leads to dangerous last-gasp strategies such as geo-engineering, nuclear and carbon markets.' Third, climate justice is supported by a widespread civil society movement launched in 2007 at the Bali negotiations, but which subsequently came to include at least one Latin American government, Bolivia.

Feminists working on climate change are connecting the dots between these various oppressions, to warn how, in times of crisis, their opponents are emboldened. In a report, 'Looking Both Ways,' the group Asian Communities for Reproductive Justice document Hurricane Katrina's deeper political damage: 'Following a disaster, women of colour – particularly African American women, low-income women and immigrant women – are routinely targeted as burdens of the state and the cause of over population, environmental degradation, poverty, crime and economic instability' (Asian Communities for Reproductive Justice, 2009). And more than for men, ending women's economic instability is a vital component of the struggle for justice.

Ecofeminist/eco-socialist networking for climate justice

It has never been more important to draw together ecofeminist and eco-socialist insights to link issues, analyses, challenges and alliance-building efforts. Is eco-socialism on the agenda? The short answer is no, there are not sufficient cadres and networks of organizations capable of connecting the dots, globally or in any national setting. But such a movement will have to emerge if we are to survive, so the following are some ways that this challenge might be viewed.

Red-green organizations are networking and expressing eco-socialist sentiments

The components of the eco-socialist movement are thousands of organizations in all parts of the world whose own assumptions about fighting environmental degradation are increasingly anti-capitalist. This is abundantly evident from the manifestos, analyses, press releases, demands, leaflets, slogans and other expressions of voice that they have generated in recent years.

The networks are typically single issue and do not sufficiently link across subsectors of environmental justice

The biggest intellectual problem these movements face is linking their concerns across other sectors. This is often because the networks come together around specific targets, and because their funding sources or major in-house intellectual resources are extremely deep within the single issue they address, but unable to move beyond it.

The networks' analysis is sometimes delimited by the specific problem they are addressing

As single issue networks, the organizations generally view the attack by capitalism on nature as a problem that they may not be in a position to name, much less propose sweeping large-scale solutions to. That has generated a void, not only insofar as naming the problem (an environmentally voracious capitalism), but also naming a global-scale socialist solution – with, of course, profound respect for difference and the uneven development of both capitalism and the movements against it. A further problem is that most such manifestos by these movements have not been particularly conscious of gender. And finally, the other kinds of interlocking and overlapping oppressions and resistances – along lines of race, indigenous heritage, different ableness, sexual preference, generation and other divisions – are not sufficiently respected to generate a strong critique.

The networks hunger to continue building links

The obvious next step for groups like Climate Justice Now! is to make common cause with other movements addressing environmental issues where similar analysis, strategies, tactics, enemies and allies can be found. There is a huge gap, though, in information about each other, since with a few exceptions (for example, the World Social Forum or protests at major world summits or meetings of well-networked organizations such as Friends of the Earth International), these organizations have no opportunities to get together in a systematic way. And yet it is inexorable that these links will become ever stronger, and it seems inexorable, too, that to properly address the challenge we face, nothing less than planetary and species survival, we encounter the overarching power of patriarchal, racially divisive, uneven capitalist 'development.' Climate justice will require nothing less.

4 Kilimanjaro Initiative

Working with youth in Nairobi

Sadique Bilal Issa

Mount Kilimanjaro provides an interesting symbol for climate change in East Africa because the loss of snow at its peak provides stark evidence that the climate is changing and water is becoming scarcer. The organization I work for, the Kilimanjaro Initiative (KI), links youth in Kibera—Nairobi's largest slum—with river protection and leadership development, using Mount Kilimanjaro as our focus. And our work also addresses climate change—though it started as a way of fostering alternatives to youth violence.

KI was founded by a victim of gun violence in Nairobi named Tim Challen. His shooting, and survival, prompted him to deeply consider what it is that can bring a young person to shoot someone else, and led him to seek solutions to youth violence through community-based organizing and leadership development programs in Kibera, Nairobi. Some facts about Kibera: one million people live in an area of five square kilometres; there is only one primary school; there are very few communal spaces where children and youth can play, and these public spaces are often nearly deserted because they are used for criminal activity. In Nairobi as a whole, there are about 110 informal settlements where basic services are lacking, and the wide gulf between rich and poor seems nearly unbridgeable.

KI is founded on the basic principle of listening to youth, ensuring youth are the driving force, and making sure all stakeholders are involved in discussions. Every year KI brings youth and other stakeholders together to discuss why youth get involved in criminal activity, and to develop alternatives to violence for and with young people. Local knowledge and local power make things run.

Where you want a community to develop, look to youth because they have energy and willpower to move things forward! KI supports youth as advocates for constructing change. But particular youth groups have to show KI how they will engage the community before KI will engage with them.

KI uses sports as a way to raise global awareness on urban safety, and to support youth as advocates for constructing change. Every year, we take ten youth from around the world for a two-week leadership training, and then the youth climb Mount Kilimanjaro, raising funds from donors as they do so. In previous years, our climbs have included youth from South Africa, Brazil, Burundi, Kenya, and Tanzania. In 2011, the KI Leadership Training theme was "Green Economy, Sustainable Development, and Greater Urban Safety," and we added a long-distance bicycle ride to our climb. Our funding comes from other big non-governmental organizations (NGOs), personal donations, and global membership fees.

We also organize sports events in the slums: cycling, football matches and other sports competitions. We are developing more sports fields in Nairobi and other low-income areas in East Africa, and we are also developing community centres—giving local people a place for meetings and communal planning. We always ensure that youth are the driving force in this process. We use a sports-community model where local/village community committees spearhead everything, help to identify problems, and come up with realistic possible solutions. The sports field committee helps to coordinate activities, mobilize communities around projects and activities, and organize the necessary work. KI is trying to upgrade or create new sports fields, and build up groups of youth around the sports fields to maintain the fields.

Related to climate change, there are issues with flooding of our first Kibera slum sports field because of its proximity to the Nairobi Dam, so we are trying to build a proper drainage system around the field so that water can drain. KI talked to the community and those groups of youth that could be prone to crime—young people who were previously engaging in criminal activities—and because of our joint initiatives they are now reformed and doing constructive community work. These youth did the manual work and they dug the drainage ditch around the field. So, the sports field, which everyone uses, is the result of their hard work (Figures 4.1 and 4.2).

Figure 4.1 Sports field located 50 meters from the Nairobi Dam, in Silanga village, part of the huge Kibera slum in Nairobi. The sports field was upgraded by KI volunteers in 2011.

Photo credit: Sadique Bilal Issa.

The water behind the Nairobi Dam is polluted from industrial effluent, raw sewage from blocked, broken and overloaded sewers and from Kibera slum, and all types of garbage. Polluted waste from the surrounding settlements has contributed to silting-up behind the dam, and it is now full of water hyacinth and other plants so that it appears more like a marsh. Heavy rainfall during storms, which are increasing in irregularity and intensity due to climate change, causes serious flooding throughout the catchment and around the dam. Since there is no clear outlet, the water spills back into the surrounding areas, where people may have built structures during the dry season. In the absence of early warning systems in the surrounding poor communities, the flooding causes

Figure 4.2 Upgrades to the Silanga sports field include extensive drainage ditches as well as earthen benches for spectators.

Photo credit: Sadique Bilal Issa.

displacement, property loss and sometimes, even death for local residents. The incidence of vector- and water-borne diseases also increases during floods.

The more sports fields we can build near the river where it floods, the fewer people's houses will get inundated or washed away. We are working with the city government to resettle people from flood-prone areas so that we can create and manage more public spaces for recreation (Figures 4.3 and 4.4).

Many of the young people in Kibera are there due to climate change, since people all over Kenya are leaving their farms because of droughts, so there is lots of rural–urban migration and most migrants are moving into the informal settlements where they can survive while looking for greener pastures. We bring awareness to the youth about climate

Figure 4.3 Houses built along the Mbagathi/Ngong River in Kibera, Nairobi. When it rains, they are submerged and/or undercut by the flooding.

Photo credit: Sadique Bilal Issa.

change and how it affects them; local problems require local solutions. We include awareness of these events and the environment in our work, and we use radio stations to disseminate information.

The KI model involves KI itself, communities, academia, and students. We are already working with planning students from the University of Nairobi, and have also collaborated on water and sanitation research through the University of Denver and the Rotary Club of Denver, which provided the funding for building model community toilets.

So, Kibera youth are now working for rivers in many ways and with many partners, through the Kilimanjaro Initiative.

Figure 4.4 Residents of some houses along the Mbagathi/Ngong River have built strong concrete walls in an attempt to protect themselves against the frequent flooding.

Photo credit: Sadique Bilal Issa.

5 Climate change in South Durban, South Africa

Lushendrie Naidu

South Durban is a significant place for environmental education and action because it is home to South Africa's largest chemical industry cluster. There are two major oil refineries (SAPREF, owned by BP and Shell, and Engen, owned by the Malaysian oil company Petronas). The AECI chemical complex and two of the largest paper mills in the country are located in this area.

The South Durban Community Environmental Alliance (SDCEA) has historically addressed health issues that have arisen from the presence of these industrial sites. Research has shown that there are many health risks from the level of industrial pollution in the area. A startling thirty-two per cent of children in the area have some form of asthma, with Settlers Primary School, located near Engen, having the highest incidence at fifty-two per cent, which is the highest level of asthma in the world for children.

Durban is home to nearly five million people, and South Durban has a population of about 350,000. During the apartheid era, poor people of colour were deliberately settled in segregated areas (black, Indian, and coloured) alongside industry, unaware of the health impacts this would cause their families; while the white upper class lived further away. This housing pattern has generally been continued in more recent years, although richer people have tended to move away from South Durban entirely.

Climate change is increasing extreme weather events in South Durban, which poses new challenges for environmental education in the area. The SDCEA thus works on a number of climate change-related initiatives.

In February of 2009 we produced a book for the average citizen about climate change in the area, called *Climate change for the people of South Durban* (researched and compiled by Lisa Ramsay), which gave an explanation of climate change that was easy to read and understand. This book was distributed to schools, communities, non-governmental organizations, and others in the local area. SDCEA conducted advocacy and awareness-raising leading up to COP 15 in Copenhagen in 2009, including protests and political action. This process was continued with an emphasis on local problems and strategies in preparation for COP 16 and COP 17, which was held in Durban. The SDCEA has facilitated workshops with children in primary and high schools, government groups, and numerous community groups on climate change and its impacts in Durban.

Much of the discussion about climate change at the national and international levels is largely based on science discourse, whereas our book and workshops focus on grassroots stories which students and people in the community can more readily relate to. For example, our book included local stories about pollution and water problems (Box 5.1).

Box 5.1 **Stories from South Durban**

Clarabelle Mthembu

Clarabelle began suffering from chronic asthma when she was seven years old and living in Clairwood with her mother and three siblings. In 1966 her family was forced to move to a single room in Lamontville under the Group Areas Act. They later moved to J section in Umlazi and there her symptoms continued. When her asthma was especially bad, her mother would take her to a doctor, but Clarabelle was treated only with cough medication.

After attending meetings about health and the environment as an adult, Clarabelle connected the onset of her asthma problem to living in Clairwood near the oil refineries: "It's only now that I'm grown that I've been going to these workshops, I'm taking note of the pollution," she says. "I realize that's where it all started, but, at that time we did not know."

Patience Green

Patience Green also offers us a story of the poor health and suffering associated with life in South Durban. Ten years ago,

Patience's son Fazel died of leukemia at the age of 14. Although doctors avoided saying what had caused his cancer, Patience observed that other children living close to the refineries were diagnosed with the same problem. Patience's main concern now is for her other son and her two year old grandson who both suffer from chronic respiratory problems.

Siga Govender

Siga Govender is one of 14 local farmers who plant crops on the 180 hectares of farmland near the former site of the Durban International Airport that was allocated to them under the Tricameral Parliament during apartheid.

Siga once grew marigolds two to three times per year to support local cultural festivals, but he stopped when heavy rains made the marigolds turn black and the crop became too much of a financial risk. A heavy storm in March 2007 and additional flooding in March and June 2008 left his crops underwater and the homes of his labourers badly damaged. The land was submerged for extended periods and crucial soil nutrients were leached out, forcing him to bear the extra costs of labour and fertilizer to replace what was lost.

Excerpted from: *Climate Change for the People of South Durban* (book researched and compiled by Lisa Ramsay for the SDCEA, "Touching Tomorrow" series, edition 3, 2009, chapter 8, pages 19–24).

Climate change is bringing about changes in weather that affect South Durban's coastal environment. In March 2007, a huge flood happened in the area and November 2008 also brought devastating storms. There is an informal settlement in South Durban where destruction from these storms was highest. Changes also include increased temperatures, flooding, and drought, which pose agricultural and food security challenges such as changes in food crops, loss of crops, and reduced access to water. Access to fishery resources and the capacity of small-scale fisheries is being reduced; changes in fish populations and migration are being noted.

One example of climate change impacts in South Durban involves urban farmers who are farming on land that belongs to the former

Durban airport, which was moved in 2009 to a new site on the other side of the city. The farmers were using water from a canal for irrigation, but this canal dried up in 2010 so they began to draw water from a borehole, and because of declining rainfall they have needed to do this earlier and earlier in the year. The farmers are also starting to grow crops which require less water, but these are not major food crops (e.g. parsley, thyme, and coriander), so they are not getting much return from their work. This means even more challenges for food security. They used to grow beans, carrots, cauliflower, and spinach. There is also lots of flooding at unusual times, which compromises newly sown seeds. SDCEA provides community empowerment to farmers through education workshops, and we visit farms to discuss practices and changes over the years.

Another example of climate change's impacts is shown by the story of a fisherman named Mohammed. Over a ten-year period fish populations have been changing in terms of quantity and location of fish species. Mohammed is starting to fish in the evenings rather than in the morning. This is exacerbating conflict with government as fishermen have been denied access to the harbour and piers where they have traditionally fished under security legislation. This, together with climate impacts means they cannot catch enough fish to make a living.

SDCEA believes people have the power to change and influence their local government, and also to influence processes further up in government where voices are fewer. We are providing community-based leadership to challenge governance and enact change, which is important because of the inherent power system that exists around climate decisions. Decisions made by a select few in the West affect millions of individuals.

6 Bringing together community and academic knowledge

The Eduardo Mondlane University Faculty of Education Environmental Education Club

Eugenia Cossa

In 2010 a new graduate program in environmental education, the LEA (Licenciatura de Educação Ambiental) was introduced in the Faculty of Education at Eduardo Mondlane University (UEM) in Maputo, Mozambique.

Based on reflective and critical learning, in line with international norms for interdisciplinary environmental education, this diploma program stimulates and develops an integrated and context-based learning experience for students.

LEA graduates are expected to become agents of intervention and change, following the general principles of curiosity, autonomy, inquisitive spirit, flexibility, openness to innovation, conflict resolution and creativity.

Eduardo Mondlane University was established in 1962 as the largest public institution for post-secondary education in Mozambique. Today, it is one of 13 public institutions in the country.

The student population grew from 2,400 in 1962 to 7,000 in 2002. In 2011, the student population reached 29,929: 27,209 undergraduate students, 1,690 masters students and 30 doctoral students. In 2006, UEM offered 23 academic degrees in 12 faculties and in 2010, 71 graduate degrees and 32 postgraduate degrees, including two doctoral degrees.

There are now 168 students enrolled in the Environmental Education Programme (LEA).

Graduate Degree in Environmental Education

The graduate degree in environmental education is offered by the Faculty of Education. Its purpose is to develop professionals and specialists in the field of environmental education who are able to:

- Integrate environmental issues into school curricula.
- Conceive, monitor, evaluate and manage projects and programs in environmental education towards sustainable development.
- Promote and conduct research in environmental education.
- Promote socio-environmental awareness and social and environmentally sustainable development.
- Promote an understanding of national environmental policies and laws in order to encourage a critical analysis of the relationships between society, education and the environment.

The Environmental Education program is four years long, or eight semesters. In addition to mandatory curricular activities, the Faculty promotes extracurricular activities which generate an academic debate based on real problems facing the country of Mozambique and, in particular, its capital city of Maputo.

Through these activities, we hope that students will gain a context for their pro-environmental actions and increase their commitment to environmental education and to the environment itself.

This initiative has been well received by the Faculty, which motivated us to create a club: the Faculty of Education Environmental Education Club (Clube de Educação Ambiental da Faculdade de Educação—CEAFE).

The Club's general objectives are to sensitize, coalesce and unite individuals and/or organized groups interested in promoting creative practices related to environmental education and the environment; and to encourage and promote innovative solutions and other alternatives for environmental issues.

Its specific objectives include creating and promoting environmental awareness for the development of sustainable practices for the use of natural resources; promoting good practices, habits and behaviours to find solutions to environmental issues; encouraging communities to find solutions to environmental issues of local and national import and relevance; and promoting multidisciplinary understanding and insight on environmental problems and issues.

The Environmental Education Club was created in response to the following objectives:

- To create and foment awareness of the environment, sustainable development and sustainable use of natural resources.
- To spread awareness to various groups.
- To understand environmental issues using a multidisciplinary approach.
- To develop skills and abilities which will aid in the development of solutions and contribute to the resolution of environmental challenges.
- To promote best-practices and habits.
- To encourage communities to look for solutions to environmental problems at the national and local level.

The following activities have been developed and are being implemented in order to accomplish the above-mentioned objectives:

- Promote knowledge and interdisciplinary discussion of environmental issues within academia, as well as an understanding of the complexities of this subject.
- Disseminate new concepts and methodologies on environmental and resource management at the academic level.
- Organize clean-ups and environmental stewardship campaigns, actively involving students, faculty and staff.
- Disseminate environmentally correct practices, through an environmental education newspaper/newsletter.
- Develop a new understanding of the environmental within academia and actively participate in the improvement of waste management on campus.

The Environmental Education Club and Climate Change Adaptation

Through the Climate Change Adaptation in Africa program, a joint initiative of Canada's International Development Research Centre (IDRC) and the United Kingdom's Department for International

Development (DFID), the Environmental Education Club has developed projects and conducted research in urban and peri-urban communities in the province of Maputo with the objective of understanding the mechanisms for climate change adaptation. The Club is currently implementing a project entitled Environmental Education at Eduardo Mondlane University. This project is supported by the Faculty of Education and aims at developing social models for promoting environmentally-conscious behaviour within academia.

The project is also aimed at encouraging students to look for solutions to climate change within their environment and with the help of their peers. As part of this project, students work with two civil society organizations (CSOs)—Women, Gender and Development (MUGEDE) and Environmental Justice (JA!). Through these internships, students gain understanding of practical, hands-on solutions to environmental and resource management and in turn, help to disseminate pedagogical information and methodologies for environmental education in local communities.

The Club's initiatives in its first year included beach clean-ups and tree planting on the university campus. We hope that the Club will contribute to the production and dissemination of knowledge regarding environmental education; create specific programs to support extracurricular activities; encourage the exchange of ideas, experiences and projects in favour of environmental education and the environment; energize intercultural and institutional actions which may improve international collaboration for environmental defence; and promote and support activities which contribute to environmental stewardship and strengthen the concept of nature as part of our heritage.

According to CEAFE's by-laws, anyone can join CEAFE who voluntarily wishes to become a member and who relates to environmental issues, in any form. The membership also includes the Director of the Faculty of Education, the Head of Department who manages the LEA program, LEA Course Directors, and all LEA students.

This initiative has been well received by the Faculty of Education because of the opportunity it provides students to strengthen their self-esteem and to foster a spirit of citizenship from the very start of their professional lives.

The idea of this environmental club is to give students the opportunity to link theory and practice. Through practical work outside the university, students can experience for themselves the crucial importance and value of combining community knowledge and academic knowledge.

With reference to climate change adaptation in Mozambique, the CEAFE is an organization with plenty of potential for generating effective and appropriate responses to climate change, since it is largely composed of educated, idealistic and motivated young people who have an interest in practical solutions and a great deal to offer.

7 Community river restoration in South Africa

Simphiwe Nojiyeza

The Centre for Civil Society (CCS) at the University of KwaZulu-Natal is an academic unit with a great many community connections. It runs several environmental justice research projects related to climate change, and supports community scholars in South Durban and North Durban. The CCS works with protest groups, unions, and social movements in critiquing carbon trading at local, national, and international levels. Another CCS project also is working in partnership with the University of Witwatersrand's Centre for Social and Legal Studies (CALS) units to support opposition to the prepaid water meters which are currently being introduced in many parts of South Africa. CCS partner organizations have direct experience in communities which are developing adaptation strategies.

Besides my academic studies at CCS, I also work at WET Africa—Waterways Environmental Transformation—a non-governmental organization (NGO) which has worked on watershed improvements in South Africa since 1998. Its projects mainly focus on restoring rivers which have been devastated by industrial pollution and wastewater from treatment plants, which release E. coli-laden effluents (Lin and Biyela, 2005: 258). To address this, WET Africa rehabilitates rivers, ensuring that there is natural vegetation to restore the health of the riverbed. The quality of water is tested using the South African Scoring System (SASS), a rapid river health monitoring protocol, and the Indicators of Hydrological Alteration (IHA) (Dickens and Graham, 2002: 1: Taylor, Schulze and Jewitt, 2003: 1). The SASS is used as a routine bio-monitoring tool for in-stream flow, as a principal bio-monitoring tool for the national river health programme, and as a tool for environmental management, assessing the ecological state of aquatic ecosystems, objectives for rivers, emerging problems, changes in the ecosystem due to development, and compliance monitoring (Graham, Dickens and Taylor, 2004: 27; Dickens and Graham, 2002: 2). The flow requirements of rivers in South Africa, and the ecological reserve within a river system as determined/required by the National River Health Programme and the Water Act of 1998, are also assessed using the SASS.

WET Africa also addresses climate change legislation, in light of the protections in the South African constitution (Sections 24 and 27), which state that every South African has a right to a healthy environment and to food security, both of which are threatened by climate change. The South African Water Act of 1998 highlights issues related to climate change adaptation such as ecological reserves, water flow, and the consequences of not observing the spirit of the constitution and water act. The National Climate Change Response Strategy for South Africa (2004) puts forth strategies for addressing climate change in South Africa.

For instance, eThekwini (Durban) Municipality has crafted the Headline Climate Change Adaptation Strategy and Municipal Adaptation Plans for the Health, Water and Disaster Management sectors. In addition, it initiated community adaptation projects in the Ntshongweni, Buffelsdraai, and Ntuzuma areas; facilitated the establishment of the Durban Climate Change Partnership; and developed its "Energy Efficiency Guideline," "Sustainable Waste Management Guideline," and "Guideline for Designing Green Roof Habitats." All of these initiatives depend on the involvement of civil society groups to monitor these plans and guidelines. However, there are only a small number of Civil Society Organizations engaging government on water resource management, governance, and climate change adaptation. The municipality's focus remains on water supply and, to some extent, climate change mitigation (Dickens and Graham, 2002: 1; Roberts, 2008: 529; eThekwini Municipality, 2009; eThekwini Municipality, 2006; eThekwini Municipality, 2010a, 2010b, 2010c).

WET Africa provides knowledge, resources, and skills on waterway and river restoration and waste separation and recycling at source. It attempts to bring water management structures, such as catchment agencies, up-to-date with adaptation priorities. This involves critiquing Catchment Management Forums (CMFs) where participants participate voluntarily but have no real power to determine what happens. As part of the national water resources strategy and observing the National Water Act of 1998, South African River and Catchment

Systems are managed in a voluntary capacity by CMFs. Originally there were supposed to be 19 statutory Catchment Management Agencies (CMAs) established to govern and manage water resources in South Africa. The number of CMAs is now reduced to nine, and only two CMAs are now operating in South Africa. In KwaZulu-Natal where Durban is situated, only one CMA is going to be established. Local authorities, such as the eThekwini Water and Sanitation Department, are anticipated to play an important role in such structures.

WET Africa also does research on water conservation strategies, rain water harvesting, the reuse of grey water, reforestation, the impacts of floods and waste, flood control, crop rotation as a climate change adaptation strategy, and disaster preparedness in the face of flooding and drought (Fairbanks and Benn, 2000: 240, Whitfield and Taylor, 2009: 838). We have contributed research on urine diversion (UD) toilets, which save 12 litres of water from each toilet flush, although there are a lot of equity and social questions that still need to be resolved (see Nojiyeza and Galvin, 2011) (see Figures 7.1 and 7.2).

The watersheds where WET Africa works in North and South Durban include those of the Umhlangane River, Piesang River, Umlazi River, Umsunduze River, and Umngeni River (see Figure 7.3). WET Africa organizes river maintenance teams to remove garbage, plant native vegetation, and monitor and improve water quality, all in alliance with local farmers who rely on the rivers to grow their crops. The mission of WET Africa is to restore rivers from source to sea; to protect and enhance human well-being in the face of climate change and vulnerability; and also to establish various river restoration initiatives in each of the watersheds where we work.

We do this by providing examples of local adaptation methods and initiatives, which are implemented, managed, and made possible by the involvement and participation of community members, rather than being large-scale top-down mitigation strategies implemented by the government. Climate change discussions often happen between governments and policy makers at high-profile international events such as the United Nations Conference Of the Parties, while local communities are excluded, though they bear the brunt of climate change. In contrast, a way of getting the community involved and mobilized around these issues is to organize communities to take ownership of their water (see Figure 7.4). This can include taking action to reduce solid waste and effluents from industry and reduce poor sanitation. WET Africa organizes

Figure 7.1 Urine diversion (UD) toilets separate liquid from solid waste to speed the drying of solids for safe use as fertilizer.

Photo credit: Patricia E. Perkins.

Figure 7.2 UD outhouse in Umzinyathi, northwest of Durban. Piped water and sewer services do not extend to this area.

Photo credit: Patricia E. Perkins.

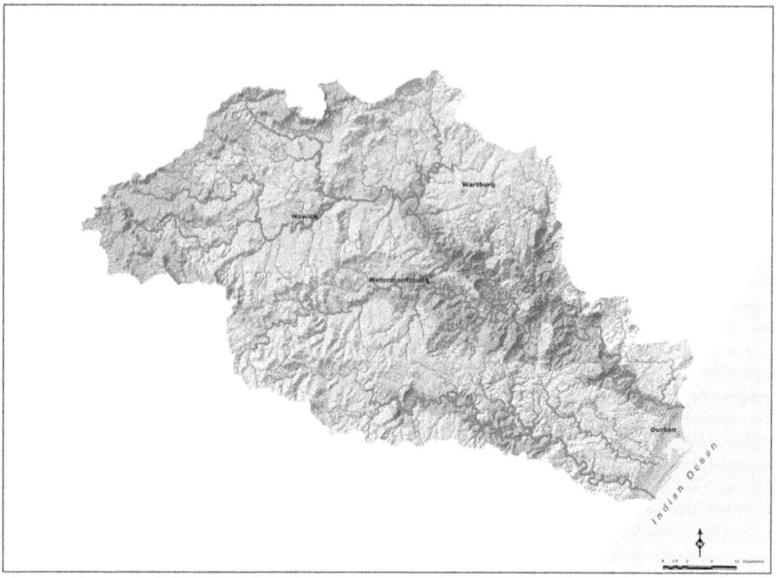

Figure 7.3 Durban's water supply. This map shows Durban's principal freshwater sources, the Umngeni River (with its main tributary, the Umsunduze River), which enters the Indian Ocean on the north side of Durban, and the Umlazi River which is just south of the city.

Source: Department of Water Affairs and Forestry *et al.* (2002). *State of Rivers Report—Umngeni River and Neighbouring Rivers and Streams*, p. 6. Available: http://www.dwaf.gov.za/iwqs/rhp/state_of_rivers/state_of_umngeni_02/umngeni_02.pdf (12 October 2012).

river clean-ups with community members. The way we work is to mobilize a team of about 100 people from the community who are willing to get involved in removing the solid waste and restoring the health of local rivers. Before rehabilitation projects begin, environmental impact assessments are conducted and the water quality is tested. The residents are given masks, waterproof boots, waders, life jackets, and gloves to protect them from potential hazardous waste; they are taught how to be careful around the water, and given training on water safety. Then we set to removing the waste so that vegetation can grow. There is also a social enterprise aspect to this work: the waste that is removed is bought by recycling and plastics firms and/or given to the communities, to help create jobs and support local

Figure 7.4 Hand pump for water provision in Umzinyathi.

Photo credit: Patricia E. Perkins.

economic development (Dickens and Graham, 2002: 3). WET Africa also educates communities on flooding, drought, and other climate change impacts (Dickens and Graham, 2002: 2). We have managed to restore several rivers in South Africa and are starting to mobilize throughout KwaZulu-Natal.

WET Africa received funding originally from the Clinton Foundation, and now communities also raise funds through the sale and recycling of waste. River Health committees are established by people who live within a catchment area. They become river monitors and come to WET Africa for support on clean-up projects (what equipment they need, how to organize and provide training, etc.) Environmental education is a critical component, since the aim is to ensure that once a river is cleaned up, the garbage does not return. This happens through community-led education workshops.

The Isipingo river in South Durban was named by the minister of water affairs in September 2010 as an example of best practices in the country, thanks to the work that WET Africa has been doing to rehabilitate that watershed. This is part of an "adopt a river" campaign spearheaded by the South African Department of Water and Environmental Affairs. River clean-up activities can improve public policy, since community members are participating, mobilizing, learning about climate change and their environmental and political rights, pushing for improved waste and water quality policies and enforcement, and engaging with local and provincial actors including government officials, NGOs, and other community groups (Lin, *et al.*, 2004: 17).

The Department of Water Affairs (DWA) of South Africa conducts audits on the performance of wastewater treatment plants run by all water authorities and service providers in the country. The criteria DWA uses in assessing water authorities and service providers include process control, maintenance, management skill, monitoring programme efficiency, credibility of waste water sample analysis, regular submission of waste water quality results, waste water quality compliance, waste water failures response management, and waste water treatment works capacity (DWA, 2009: 16). Water authorities and service providers that meet the criteria and achieve an overall score of above 90% are awarded a "green drop" certificate for that particular plant (DWA, 2009: 6; De Villiers, 2005: 196).

One of our goals is to ensure that communities understand how they can hold their municipalities accountable for water quality and wastewater treatment plant monitoring. In Durban, for example, out of 27 wastewater treatment plants, only 11 were "green drop" certified in 2009: the Amanzimtoti, Isipingo/Umlazi, Umkhomazi, Craigieburn, Southern, Phoenix, Central, Umdloti, Umhlanga, Verulam and Umhlantuzana wastewater treatment systems. In 2011, the number of waste water treatment facilities that were "green drop" certified was reduced to nine when the Isipingo and Verulam plants lost their previous status (DWA, 2009: 43; DWA, 2011: 21).

In terms of monitoring the quality of drinking water, the DWA established the drinking water quality regulation programme in 2005. Its objective is to ensure the improvement of tap water quality by monitoring compliance by water authorities and providers, who are evaluated on the following: water safety plans, asset management and drinking water quality performance publication, microbiological and chemical compliance. A water authority and provider is awarded a score for water safety plan, process control and maintenance competency, efficiency of drinking water quality monitoring programme, credibility of drinking water sample analysis, regular submission of drinking water quality results to DWA, drinking water compliance with the South African National Standard (SANS 241), drinking water quality failure response management, responsible publication of drinking water quality management performance, and efficacy of basic drinking water quality asset management. In both 2010 and 2011 eThekwini Municipality managed to meet the criteria and was awarded a "blue drop" certificate. The role of non-state actors is crucial in holding government accountable through participation in CMFs like the Lower Umngeni CMF, Inanda and Naggle Dam CMFs and the proposed Umvoti-Umzimkhulu Catchment Management Agency that is anticipated to take over the governance of water resources within this catchment area (DWA, 2010: 10; DWA, 2011; De Villiers, 2005: 196).

WET Africa works with a team at each municipal level, and has maps that show the state of quality in each and every river in the country, to determine which rivers are the most vulnerable and require attention most urgently. The watershed approach allows us to focus on tributaries because they feed into the larger rivers, which affect the mangroves at the river's mouth. We also report waste and dumping to the authorities, since waste is sometimes dumped in truckloads by industry, and local communities which are in a position to know about these violations need to be vigilant and have good communication channels with government agents. This is also done while calling for wetlands in most of the tributaries, floodplains and sustainability of the ecosystem to be restored (Grenfell, Ellery and Grenfell, 2008: 2029). In the case of North and South Durban, communities are dealing mainly with industrial pollution, waste reduction plants that are failing, and personal/household waste, all of which require improvements in intervention strategies.

My own academic research priorities have been informed and influenced by my work with WET Africa. The organization appreciates involvement by students and researchers who can help with organizing, policy, community education, and documentation activities. WET Africa has great potential for continuing to organize community responses to climate change in South Africa.

8 Gender and climate change in Mozambique

Saquina Mucavele

MUGEDE (Women, Gender and Development – Mulher, Género e Desenvolvimento) is a non-profit civil society organization, which was established in Mozambique in 2004. Its mission is to develop active citizenship for the preservation of environmental values, for rural development and the promotion of gender equality driven by justice, tolerance, solidarity and equity.

MUGEDE is based in Maputo, the capital of Mozambique, and it also operates in several other provinces of the country. Its activists are women and young adults who have received formal and informal environmental education.

Keeping Mozambique's vulnerability to climate change in mind, MUGEDE develops activities to educate civil society, especially women, in how to care for the environment to prevent and mitigate the effects of climate change.

Historically, Mozambique is the country most affected by natural disasters in Southern Africa. Mozambique's long coast is situated in one of the most active basins for tropical cyclones, the Southwest Indian Ocean. According to the EM-DAT International Disaster Database (http://www.preventionweb.net/english/countries/statistics/?cid=117), more than 23 million Mozambicans were affected by natural calamities between 1980 and 2010.

Mozambique is especially vulnerable to climate change due to the fragility of some relevant sectors (such as agriculture, health, transportation) and the limited human, institutional and financial ability to anticipate and respond, both directly and indirectly, to the effects of climate change.

This incapacity to adapt and respond to climate change was dramatically demonstrated during the floods of 2000. A succession of tropical storms that year caused the simultaneous flooding of the Limpopo, Incomati, Umbeluzi, Save, Buze and Pungoe rivers. More than 4.5 million people, a quarter of Mozambique's population, were affected and about 800 people died.

Since 1980 Mozambique has also faced eight major droughts, due to irregular and unpredictable precipitation. The rainy season may not begin when expected, resulting in erratic seed planting periods. The rains may be short and very intense, causing soil degradation and superficial drainage/runoff, or insufficient, causing crop failures.

In Mozambique, droughts are caused not only by a shortage of precipitation, but also by reduced water flows originating from dams in neighbouring countries. Mozambique is part of nine international hydrographic basins, so the whole country is vulnerable to both floods and droughts related to changes in water releases from upstream dams.

Eighty per cent of Mozambique's population lives in rural areas and depends on agriculture for their survival. Approximately 70 per cent of energy consumption in Mozambique is produced from biofuels. Approximately 18 million cubic metres of forest are destroyed annually for the production of firewood and charcoal. Even in the city of Maputo, firewood and charcoal are the major sources of energy for cooking and heating, especially in the informal or poorer neighbourhoods. These fuels are in short supply.

Water supply, too, is problematic in the city, and only about 35 per cent of Maputo's population has access to piped water (either in their house or via a public standpipe).

Because of their gendered work and household responsibilities for finding fuel and water, cooking, childcare, care of the sick, and cleaning, women are the catalyst for environmental intervention and consequently, for climate change adaptation.

The direct repercussions of drought, floods, cyclones and the HIV/AIDS epidemic affect women's lives and are closely related to climate change. For example, when food scarcity occurs (due to crop failure or death of livestock brought about by floods or drought), young girls may no longer be able to attend school, since they may need to spend more time on household responsibilities. Women's crucial role in household water management and gathering of firewood means they bear the brunt of climate change.

MUGEDE's strategies for climate change adaptation and mitigation

MUGEDE's work involves a wide range of community development and education activities related to climate change adaptation and awareness.

Through modest funding, MUGEDE has developed initiatives to promote environmental education in schools as well as in rural and urban communities.

Slowly, one can start to notice a slight change in society's behaviour towards the environment. Mozambique is developing a national climate change strategy, including an action plan, with civil society consultation. In partnership with the non-governmental organization OCPA (Observatory of Cultural Policies in Africa – Observatório de Políticas Culturais em África) and the Mozambican federal Environmental Action Coordination Ministry, MICOA (Ministério para a Coordenação da Acção Ambiental), MUGEDE organized an international conference on Woman and the Environment ('A Mulher e o Meio Ambiente') in November 2010. The goal of this conference was to reflect on women's role in times of climate change and to pave the way for women to minimize poverty and increase success.

MUGEDE has started a reforestation project for communities, which produce charcoal and firewood, especially where natural resources are vulnerable to climate change.

MUGEDE promotes sustainable use of local biodiversity in both rural and urban areas; carries out capacity-building related to the environment with a focus on climate change in local communities; promotes initiatives for the expansion of sustainable energy practices and use of adequate stoves by rural women; sensitizes and informs communities about the necessity to treat water and use it sparingly; encourages conservation of water in agriculture and rural communities; educates rural women on how to care for water during drought, when cholera and diarrhoea are prevalent; teaches women about the risks associated with building their homes in areas that are vulnerable to drought and erosion; educates about the importance to use building materials that are more resistant to the effects of climate change; and helps residents to identify safe areas where they can take shelter in case of emergencies.

MUGEDE has a number of additional projects still in the proposal stage or awaiting funding, which also would help to strengthen

Mozambican women's abilities to respond effectively to climate change.

A project on Gender and Food Sovereignty (Género e Soberania Alimentar) was also created for the purpose of reducing the effects of climate change in the lives of women, particularly residents of rural areas. This project's objective is to identify sustainable environmental practices of food production, accessible techniques of production and food conservation/preservation, with a focus on the use of more stable barns as well as the cultivation of highly nutritious crops which are resistant to drought, such as beans, sweet potato and corn.

Together with Gender Climate Change and Justice, a partner non-governmental organization, MUGEDE continues to apply for grants to fund its project to work with other organizations to equip women with better tools to prevent and mitigate the effects of climate change.

Climate change is one of humanity's biggest challenges. It threatens to perpetuate poverty and hardship and for this reason, men and women must work side by side with equal rights and opportunities in order to face this challenge.

Women must be involved in discussion forums related to the mitigation of the effects of climate change, since they have a key role in the use and management of natural resources, and are generally so gravely affected by climate change. We need more capacity-building initiatives in urban and rural communities with special emphasis on climate change prevention, mitigation and adaptation. The education of women will be the driving force to achieve positive results.

9 Climate justice

Can you educate hungry people?

Erika Mendes and Tiago Esmael

In Mozambique, the population is mostly rural and people are highly dependent on rural ecosystems for their livelihoods, so the mismanagement of water resources has huge impacts, not only on rural ecosystems but on the communities that depend on them. Floods and droughts have caused great suffering in Mozambique, so an important priority should be evening out water availability from one season (and year) to the next (see Figures 9.1, 9.2 and 9.3). There are traditions of rainwater harvesting in some areas of Mozambique, but there is currently no overall strategy for improved rainwater harvesting—despite recent studies showing this could meet the country's basic water needs.

Mismanagement and misuse of water, resulting from the government's shortsighted extractive approach to water development, exacerbates social and economic injustices. The priorities of the government have been dam megaprojects, such as the huge Cahora Bassa Dam (built in the early 1970s on the Zambezi River) and the Mphanda Nkuwa Dam (approved by the Mozambican government in 2010, to be built 60 km downstream from Cahora Bassa and financed by the China Exim Bank).

These dams are for energy exploitation, but most communities surrounding the dams don't even have access to electricity, which highlights the justice aspects of these issues. Climate change has also been left out of discussions about dams in Mozambique—the focus is on immediate "progress."

Justiça Ambiental! (JA!) has been trying over the years to include climate change in discussions on such megaprojects, but for a number of reasons including political resistance, this has been very difficult.

Figure 9.1 Heavy rainfall has washed huge gullies through the Polana Caniço neighbourhood in Maputo.

Photo credit: Patricia E. Perkins.

Figure 9.2 A storm in 2010 washed away several houses in Polana Caniço, just uphill from the Indian Ocean.

Photo credit: Patricia E. Perkins.

The communities living near dams suffer floods and droughts due to the poor management of these dams. The government tries to convince the people to move away from the river, but that is not a viable choice for them because they need the water to sustain themselves (Morrissey, 2006).

According to Mozambican meteorologist Alexandre Tique:

> the combination of floods and droughts makes many communities particularly vulnerable. In times of drought people move closer to the rivers to farm because the land is more [fertile], and they settle there to be close to their land and crops. But when the floods come they don't have enough time to head for higher ground. The constant change between a situation of drought to one of flooding means these people are highly exposed.
>
> (IRIN, 2008: 1)

Figure 9.3 Stormwater drainage canals are being extended in the low-lying Maxaquene area of Maputo. Piped drinking water provision does not extend to all households; many water pipes leak and/or are tapped by those who need drinking water; sanitation is based on pit latrines; rainstorms cause flooding which can spread disease.

Photo credit: Patricia E. Perkins.

When dam construction (or climate change) displaces entire communities, this causes great problems, because often local leadership structures are ignored and the communities do not have a say about their new location. Different languages and different cultures cause political and economic difficulties for those who are forced to migrate.

Environmental education

JA! has approached the government several times about introducing environmental education into the public school system curriculum, which the government has agreed is an important priority. But things take very long to be implemented, and non-governmental organizations and external associations that try to collaborate with public institutions find it very difficult. Improving public environmental education continues to be one of our goals.

JA! has organized environmental photo exhibitions once a year with amateur and professional photographers, including one called "Water is Life!" in 2010 which invited the public to comment and participate. We have also made a number of environmental presentations in schools. And we have organized community workshops, especially in the area around Mphanda Nkuwa. We have helped to create a civil society association in the Zambezi watershed, the "Vozes do Zambeze" (Voices of the Zambezi), and we are part of the Zambezi River Basin Committee. We have made an illustrated graphic magazine on dam construction, floods in the river valley, what a dam is and the impacts it creates. We have also organized school field trips and environmental workshops on various topics in universities and schools.

Through our grassroots work as well as our participation in international networks and regional alliances in Africa, JA! is attempting to contribute to Mozambican communities' long-term capacity to face climate change—a principal development challenge.

One question we often have to confront is this: How do you educate hungry people? How can you justify teaching people about the environment, when their primary concern is day-to-day subsistence? In a recent study, partners in the Africa Climate Change Resilience Alliance (ACCRA) found that local farmers are well aware of changes in weather patterns and see them as worrisome threats to their subsistence, in combination with other interrelated stresses such as land tenure problems, human and animal diseases, drinking water access, migration pressures, and many other constraints (Arnall, 2012).

Climate change, thus, means that environmental education and community organizing are becoming ever more crucial for local people's survival. Civil society organizations can play a key role in facilitating locally-useful, politically-appropriate, and equity-enhancing interventions. The best way to do this is part of a long and very important discussion which is a major concern when people's basic needs are constantly at risk.

10 A hot climate for civil society engagement with climate change and water in Durban

Mary Galvin

Not only is the physical climate in the Durban metropolitan area (eThekwini Municipality) getting hotter due to climate change, but the relationship between the municipality and civil society is "hot" in terms of contestation. This chapter raises questions about civil society–municipal interaction around water services and issues related to climate change, drawing on the work of civil society organization (CSO) Umphilo waManzi (Water is Life).

A hot climate: civil society–government engagement

South Africa has a strong rights-based approach to governance as a result of its apartheid history. People expect government to deliver services that they were denied under apartheid and to respond to the massive inequalities that remain. How do people engage with government around their socio-economic rights?

One way is through CSOs. Historically, non-governmental organizations (NGOs) were largely white service-based organizations supporting community-based organizations (CBOs), which were driven by people on the ground doing community development and organizing. Within the broad civil society framework, differences emerged between NGOs and CBOs over funding and "who has a mandate to speak for whom." Over the past decade, social movements have arisen outside the "development" circles, allowing people a direct voice around issues that concern them and creating a structure for organizing advocacy actions.

Locating this range of CSOs along a continuum from social change or activist organizations to technical organizations clarifies their differences.

There are strong ideological differences between CSOs: most CBOs and social movements more closely embody social change organizations, and technical organizations work closely with government. NGOs are spread across the continuum, influenced by the role of funding in civil society–government relationships. After the transition to democracy in 1994, donors began to provide funds through the government rather than directly to NGOs, placing NGOs in a tight spot. Their critique of government can become cautious or muted to avoid jeopardizing their access to funding (Galvin, 2010a; Galvin, 2010b).

For example, in terms of climate change networks in South Africa, Climate Justice Now! is considered a social change organization as it is highly critical of government NGO and CBO participation, while the South African Climate Action Network is regarded as a mainstream organization that cooperates with government.

Of course how CSOs approach government is only part of the equation. We also need to ask "How does government engage with its citizens?" Civil society's role in general and in terms of water supply is well established in South African legislation and policy. The requirement to consult citizens and/or CSOs or to seek their participation is part of the legal framework for water management, including the Municipal Systems Act (2000), Strategic Framework for Water Services (2003), National Water Act (No. 36 of 1998), and National Water Services Regulation Strategy (2010).

Yet this is typically not happening. Part of the problem is a lack of appreciation for levels of participation. Participation can be conceptualized as a ladder with a bottom rung of "persuasion," where the level of participation increases through the higher rungs of information sharing and consultation. The top rung is "empowerment," which requires the most active participation of communities (Arnstein, 1969). Some issues require only information sharing or consultation, where others require full participation or empowerment.

But it is not simply a matter of officials not understanding that there are different levels of participation. The tendency is for government to simply communicate or consult citizens or CSOs in a perfunctory manner. There are consumer hotlines, mayoral road shows, and invitations to consultations on development plans that are placed in local newspapers. The present paradigm is that government officials make decisions and, if anything, arrange an open information sharing session or expect to co-opt CSOs as one member of a large committee (Galvin, 2009).

Finally, part of the problem is how issues get caught in a political tug of war, which disregards the voices of citizens and CSOs. For example, there are processes set out in law for civil society participation in managing water resources, such as Catchment Management Associations (CMAs). Progress was made in establishing the first few CMAs, but government stopped the process mid-stream while negotiating a political plan to change the number and boundaries of CMAs, causing many CSOs to become wary of participating in CMAs (Enright interview, 6 July 2010).

It is the very lack of communication and of being given a voice that has fed widespread protests in response to poor people's frustrations with government's delivery of services (Friedman, 2005). Class equity, which still overlaps powerfully with race in South Africa, is a key issue underlying people's rights to services, which climate change is likely to exacerbate further. With few, if any, effective channels to communicate with government around service delivery expectations, community frustration escalates. How might this apply to water services in the face of climate change?

A hot climate: climate change and equity issues

Water demand already outstrips supply in Durban. The government's water reconciliation plan shows that the Spring Grove dam was supposed to be built in 2009 to respond to this shortfall, but funds were evidently used to prepare for the World Cup and the dam was delayed. Although there has been heavy rainfall since 2009, this is expected to turn to dry spells soon, and then the municipality will have to resort to imposing water restrictions (Macleod interview, 2011).

The municipality would most likely introduce a restriction of 400 litres per household per day. Currently middle class families tend to use this amount of water per person, so it will be quite an adjustment for an entire family to live on that much (Macleod interview, 30 April 2010). In contrast, many poor and working class families live on the free basic water allotment of roughly 50 litres per person per day (for an estimated household size of five people). So it is the wealthy who are likely to feel the pinch of living on less water, revealing current injustices.

Water scarcity alongside the high capital cost of a new dam means that we can also expect the cost of water—and tariffs—to increase.

But how will the tariff curve be shaped? In other words, will this be passed onto users equally? Poor people who go above the free basic allocation due to a typical large family size would suffer. Or can the tariff curve be steep and concave enough that high end users bear the brunt of the increase? (Bond, 2002).

These equity issues show how the impacts of climate change in an urban African city is mediated by municipal policies. To make the bland statement that "variable rainfall will put more pressure on water services and even more pressure on poor people" misses the complex role of the municipality in adding or buffering that pressure. How can civil society engage with these issues?

One example of civil society engagement is Umphilo waManzi's work in three related areas.

Area 1: Challenging the current paradigm of participation

One of Umphilo's first projects was administering the Water Dialogues–South Africa, a national multi-stakeholder dialogue based on its research in eight areas of the country (http://www.waterdialogues.org/south-africa/index.htm). The Water Dialogues promoted civil society's inclusion in water deliberations and decisions, and the need for such participation was the basis of many of its final recommendations (Galvin, 2009). To address water struggles requires the kind of paradigm shift or change described by the Water Dialogues' stakeholders as:

> A radical rethink of *how* we work to ensure effective, equitable and sustainable delivery of water supply and sanitation provided affordably to all . . . Local and national government need to acknowledge the range of voices which must be heard if lasting improvements are to be achieved in service delivery. Discussion of the way forward to better servicing must be broadened to harness the ideas, skills, competencies and resources that exist in our communities. To pursue a developmental agenda means formulating local solutions for local realities . . .
>
> (Galvin, 2009: 7)

At the local level, Umphilo has also conducted research on how civil society engages with eThekwini Water and Sanitation (EWS), recognized as one of the top public water utilities in South Africa and even in Africa. It highlighted and assessed various approaches used by civil society to advocate change—the hydropolitics of eThekwini (Galvin, forthcoming). While civil society succeeded in initiating some change through dialogue and regular engagement with the municipality, citizens were left with quiet resistance and protest to pursue other outcomes.

For example, through his engagement with CSOs in the Water Dialogues, the EWS director decided to increase the free basic water allocation in Durban from six to nine kilolitres. He also adjusted a "citizens voice" programme that was piloted in Cape Town (see Box 10.1). It provided civic education on aspects of water and sanitation and then established "user platforms," regular meetings in which citizens raise systemic water and sanitation related problems directly with EWS officials.

Box 10.1 **"Citizen's Voice" in metropolitan areas**

The Citizen's Voice project was piloted in Cape Town in 2006 and subsequently introduced in other cities including eThekwini, Msunduzi and Ekurhuleni. Citizen's Voice provides training to community members and ward committee members on a range of water and sanitation issues. The aim is to provide civic education about the issues related to water services so that participants have a foundation to hold officials and politicians accountable for sustainable water provision and are able to articulate their problems and issues in a way that permits officials to hear them and consider them at a systemic level. Officials meet with citizens to share information, and so that citizens can raise and discuss their problems and concerns around water and sanitation with those responsible.

The project provides an excellent example of deepening citizen participation, although since there is no clear and independent driver, the process can be taken up and used by organizations for their own goals. There is a lack of oversight and quality control to ensure that this does not take place and to maintain standards of training. As the process is now endorsed in the National Water Services Regulation Strategy (2010), practitioners are working to address these weaknesses (Galvin, 2011).

However, some counsellors are threatened by the municipality meeting directly with citizens since they perceive this as undermining their role. So formally civil society meets with the municipality, but in practice EWS issues its invitation to CSOs cautiously. This limitation, along with other controversies such as those related to urine diversion (UD) toilets, have started to diminish EWS' achievement in charting a new course of civil society-municipal engagement.

Area 2: Action research—grounding people's truths

Umphilo has also been grounding issues with action research on how things work in practice, giving power to people's experiences and "stories" as much as the "data" that is often treated as the factual "truth" in the technical water sector. One example is the 80,000 UD toilets built by the municipal government in the peri-urban and peri-rural areas of eThekwini (see Box 10.2). Umphilo has found that most are not working (well) because of problems with both the social acceptability of emptying them and the lack of consultation and education before introducing them (Nojiyeza and Galvin, 2011). As part of our climate change and water work, Umphilo has been working with one community to explore a pilot project that addresses these problems and explores options for improved take-up as part of local climate change adaptation.

Box 10.2 Resisting urine diversion toilets in eThekwini

One community activist put it simply: "Black people want to experience the flush." People who have not had water/sewer services want the same level of services as exist in wealthier neighbouring areas.

When Durban's municipal boundaries were extended in 2000, EWS suddenly had tens of thousands of new households in its jurisdiction that required sanitation. Ventilated improved pit (VIP) toilets had been found difficult and highly expensive to empty. So EWS turned to dry sanitation in the form of UD toilets, a form of "ecosan" that is considered by many international sanitation specialists as the way of the future.

Why are local people practicing "everyday forms of resistance" (Scott, 1992) such as refusing to use their UD toilets, turning them into storage cupboards or transforming them to flush toilets using a French drain? Umphilo's research in six different wards consistently showed that the municipality did not consult with people on their options (partly because they really had none) and provided user "education" that was of extremely poor quality. This meant that people often had a toilet they did not want, could not use properly, and were left to empty themselves. Waste did not dry out properly, and people are angered by having to empty wet sludge.

People's non-use of UD toilets threatens to turn into a public health crisis. Umphilo believes that this situation can be averted by implementing a full community consultation and education process, and providing a municipal service to empty UD toilets.

Area 3: Advocacy—from awareness-raising and acting as a watchdog to protest

The municipality encourages water conservation, squeezing the poor to use even less water. For social equity reasons Umphilo is highlighting the need for the middle-class to use rainwater harvesting and greywater systems. It has done surveys of middle-income households' use of rainwater harvesting and produced participatory videos and posters on how using these systems can save money and water.

Umphilo also plays a watchdog role. The national Department of Water Affairs has put into place Blue Drop and Green Drop awards, which recognize municipalities for having achieved a high standard of water quality and waste-water treatment respectively (see chapter 7). Ongoing sampling takes place and results appears on a website, enabling civil society to monitor this process. Citizens can keep an eye out in their local area to ensure that what their municipality reports is actually what is happening on the ground.

Finally, Umphilo joins other CSOs in strategizing and organizing protest actions related to water services and resources. While progress can sometimes be made by engaging with government, additional pressure is needed for better water policies, such as steeper tariff increases for higher-income people and corporate water users, and other service delivery improvements.

Conclusion

Umphilo is presently working with community groups in eThekwini to identify and assert their agency on climate change adaptation and water issues. As our experiences may be applicable to other peri-urban and peri-rural municipalities in South Africa, we aim to use these findings about community level adaption to influence higher levels of decision-making in order to affect broader socio-political change.

11 Planning and climate change in Nairobi

Romanus Otieno Opiyo

Climate change is challenging the livelihoods of the poor throughout Kenya, including in Nairobi city. There is generally a low level of knowledge and awareness about climate change and its effect on water availability in Nairobi. People tend to see a very remote connection between what is happening to the environment and anthropogenic activities. The impacts of climate change in Kenya include reduced availability of and access to water, causing rationing during droughts; reduced hydropower generation (which is a major energy source for the country); and increased vulnerability of the urban poor. Due to climate variability and factors such as droughts and interference with watersheds in Kenya, hydropower has been fluctuating and it is becoming irregular. The government has introduced thermal and geothermal energy projects, among others, to address this problem, but the solutions will take some time to materialize. Those relying on electricity for their livelihoods (e.g. in factories or home-based industries) and for personal safety (requiring well-lighted public spaces) are negatively affected by these changes.

Kenya is rapidly urbanizing, with an average urban growth rate of 3.9 per cent per year for the period 2005–2010. In 2007, about 25 per cent of Kenya's population lived in cities, and according to the 2009 national census 32.3 per cent of the population were living in urban centres. This growth is largely due to a high level of migration to the cities, fuelled by rural poverty and a dwindling per capita ownership of farming and grazing land. Nairobi attracts the bulk of rural–urban migrants.

There is a lack of concern and commitment by the State agencies to building resilience, although there is emerging interest on the part of the Government of Kenya in reclamation of watersheds and rivers. In 2007, the Government of Kenya launched its vision for national development over the next 20 or so years ("Kenya's Vision 2030," Ministry of Planning and National Development). The plan, "Vision 2030," is implemented through five-year rolling plans starting in 2008 ("Kenya's First Medium Term Plan 2008–2012").

The first five-year plan sets out environmental objectives with a forest focus:

- Increasing forest, tree cover and wood production especially at farm level.
- Conserving and rehabilitating the remaining natural forest and woodlands for environmental protection and biodiversity conservation.
- Enhancing participatory forest management.
- Ensuring that the forestry sector makes a contribution to poverty reduction.

The plan also covers management of water catchment areas, and recognizes that deforestation has caused severe degradation of the main water supply areas, reducing base river flow and disrupting electricity supply.

The new Kenyan constitution (2010) recognizes access to clean and safe water in adequate quantities as an economic and social right for all—so a certain amount of political goodwill exists on water-related issues. However, while a few projects related to Nairobi River and Nairobi Dam have been launched by the government along with other partners, there is little evidence of a commitment to addressing the issue of climate change more generally.

Nairobi Dam is located right next to Kibera slum, one of the biggest slums in sub-Saharan Africa. Its waters are quite polluted, and since 2006 there has been a government project to clean up the Nairobi River which runs down from the dam. This project includes partners such as the Kenyan Ministry of the Environment, the City Council of Nairobi, the Government of Kenya, and the University of Nairobi in partnership with civil society organizations. The project has managed to reclaim and protect some sections of the river by moving some incompatible activities that encroach on the riparian reserve, notably the section near Ngara and the area near Kirinyaga road, where a number of informal activities such as motor garages were located on the riparian reserve. The major task of the project is to reclaim the entire river and control pollution emanating from commercial, industrial and residential uses adjacent to the river.

Institutional capacity and bureaucracy

The limitations of current programmes are partly due to insufficient financial resources and technology to adapt, respond to climate change and act accordingly. Other problems include high levels of bureaucracy in local institutions, and as noted above, low levels of knowledge and awareness about the impacts of climate change on the city, its water and everyday life.

There is a general lack of policies, instruments and strong institutions for regional and city management and governability. Nairobi has neither a Regional Plan, a City-wide Development Plan, nor a Risk Management Plan. When the Environmental Management and Co-ordination Act was developed in 1999, civil society groups were not consulted, and they are not actively participating in the environmental assessment process for new land-related developments. To put up a new building in Nairobi, you must get permission from the City Council of Nairobi and National Environmental Management Authority. But these institutions simply do not have the capacity to control development in the entire city. The city of 3.1 million people has fewer than 20 municipal planners. They're overworked, and that's a problem; that's one reason we see so many slums sprouting up, since they are unable to guide and control the city's physical development and expansion. Occasionally the riparian reserves and water courses are encroached upon by developers, risking the lives of the occupants of these developments, especially during rainy seasons with regard to flooding.

Another glaring and daunting task is coordination among the policies and institutions entrusted with the duty of planning and managing the metropolis. The many conflicting statutes should be harmonized so as to have a structured and value-adding approach to the management of the metropolis. The policies and institutions should also take seriously the perceptions and aspirations of all cadres of urban dwellers, from slum dwellers to the wealthy residents, and all cadres of self-employed persons, from street vendors to multi-billionaires. This is crucial so that the city's spatial framework is seen as legitimate and binding upon all. At the same time, Nairobi has a need to protect the metropolitan environment from being a major generator of pollution. We need to attempt to ensure that pictures like the one below, where some people live in the riparian reserve, are avoided (see Figure 11.1).

Local authorities have considerable authority over land use planning and water management, and thus they play a significant role in containing major contributors to urban climate imbalance—such as the transportation and energy sectors—all of which have implications for climate change.

Most urban development models in Africa have generally ignored the plight of the urban majority, such as the informal settlement dwellers and urban informal economy operators. This mistake is also visible in the formulation of global agendas and development models—as shown in the focus of the Millennium Development Goals (MDGs), where targets such as revitalization of the informal economy are missing despite the fact that the informal economy is the kingpin in

Figure 11.1 Houses built on riparian land, 10 meters from the Nairobi Dam in Silanga village, Kibera, Nairobi.

Photo credit: Sadique Bilal.

providing work for the urban poor, whose income needs to be raised in order to meet basic needs.

At the global scale also, the effort to curb climate change has always focused on the major generators of greenhouse gases at the expense of the victims, who in most cases live in the less-developed countries. This makes it hard to contain the negativity associated with climate change, since developing countries are continuing to embrace and adopt technology and skills that are being phased out in the developed world because of the environmental threats they have been associated with over time.

These shortcomings call for development of an integrative and comprehensive urban planning and development framework at all levels that embraces progressive urban governance and empowerment of the institutions charged with the planning of urban areas. The Nairobi Metropolitan Region 2030 planning framework should ensure that problems associated with uncoordinated urban growth are handled amicably, and also take into account the desires of the urban poor, who form more than 60 per cent of Nairobi's urban population and who will be worst affected by climate change, due to the nature of their activities and the less-desirable locations where they live. Sustainable urban planning and governance is no longer a preserve of professionals and bureaucrats, but must be recognized as an inclusive process, accommodating stakeholders of all kinds.

Water governance and planning

Improved water governance in Nairobi will require the involvement of state actors in enhancing adaptability in relation to service provision. For example, different stakeholders have different capacities; the Kenya meteorological department and key service providers, such as the energy and water ministries and the local authorities like the City Council of Nairobi, are all involved in various types and stages of adaptation. Their involvement is necessary in developing a better informed public on climate change issues. An information database and structures for information sharing and dissemination are also needed. This involves issues of territorial dimension, going beyond the State apparatus and embracing partnerships with civil society organizations, and the more general awareness and informed participation by civil society.

Nairobi could even develop a Water Governance Index (WGI) using criteria such as availability, management, effectiveness, accountability, participation, equity, and others.

University–stakeholder interaction methodology

The University of Nairobi has community outreach programmes dating back to the 1960s, with departments such as agriculture and medicine being the pioneers of this noble idea, which has now spread to other faculties such as the School of the Built Environment (SBE). Students work in and with communities to share their common interests and also to appreciate how community organizations conduct their work. In the urban planning department, through various rural, urban and regional studio courses, staff and students interact with citizens and development stakeholders including government officers and non-governmental organizations to identify and map issues affecting planning and development, such as resource and governance issues. Through these studios, the department also helps to identify areas needing capacity building and improvements in interaction and communication, enhancing inclusive/affirmative participation, especially of the disadvantaged.

Examples of outreach work undertaken by the University of Nairobi include:

- The establishment of benchmarks and platforms for interaction between universities and civil society organizations, in the University of Nairobi's work with civil society organizations such as Pamoja Trust on slum improvement projects in the Mathare and Mukuru areas of Nairobi.
- The Department of Urban and Regional Planning, through its Urban Innovations Project (UIP), is undertaking wider innovative research-based outreach in the Nairobi city slums and is now focusing in two Nairobi slums, Mathare and Mukuru. They are supporting communities to prepare local plans which can inform the provision and accessibility of shared community facilities.
- Cleaning and reclamation of Nairobi River. This initiative is driven by the United Nations Environment Programme (UNEP) which formulated its "Nairobi River Basin Programme" to improve the environmental conditions of the river basin. This

exercise is ongoing, and it has brought together government agencies and the private sector.

- Participation in the development of planning, redevelopment, and conservation strategies for Nairobi River—a government priority. As one of the three rivers in the City, Nairobi River occupies the central position, and it has longstanding historical importance as it traverses many areas of the city. However, there is overt encroachment on the floodplain, neglect and apathy towards the river, with rampant pollution and diversions. A River Rejuvenation Policy is needed to spark the environmental upgrading of the river, and develop the river as an integral structuring system for the city (Government of Kenya, 2011).
- Tree-planting activities in the Karura and Mau forests with university student and staff participation.
- Combined studios between the University of Nairobi and Columbia University focusing in Ruiru municipality to address several developmental issues such as land use and development control in the area.
- Development of appropriate models of water governance that are also sensitive to climate change adaptation, focusing on the awareness and resilience of the urban poor. This is a multi-agency approach where universities including the University of Nairobi are involved, along with the UNEP, the Ministry of Environment and other ministries, the National Environment Management Authority (NEMA), the City Council of Nairobi (CCN), the Kenya Private Sector Alliance (KEPSA), Kenya Association of Residents (KARA), Muungano wa Wanavijiji (an association of slum dwellers), the Kenya Association of Manufacturers (KAM), and others. The idea of this platform is to create a framework for ensuring that all water users have access to water, but also to include other factors that may enhance living environments for the urban poor. The platform is normally active when a member of the platform has a problem with service provision. It helps in reducing the antagonistic relationship between service providers and service users.

Conclusion

Local water governance in relation to climate change still involves complex debates, despite the global concern it has merited. The governance and adaptation to climate change will require an integrated approach to build various stakeholders' confidence and capacity. We need to reduce the gap between researchers and implementers, and between civil society and government. The focus and energy should be on awareness, partnership of various actors and behavioural change.

Planning in Nairobi should be based on a true sustainable development framework which should be highly participatory and responsive to the needs and aspirations of the residents. It should be integrated with urban governance principles which enable people to appreciate both short and long-term implications of decisions taken now. This is crucial, as the poorer countries like Kenya will likely continue to suffer irreparable losses which will be heavily shouldered by the next generation.

12 Climate vulnerability and resilience in local communities

Elizabeth Lorimer

In any climate justice work, which by definition is focused on the needs of the most vulnerable people, it is important to acknowledge that the most vulnerable can also show the greatest resilience. Approaching climate change with a justice lens recognizes the effects that climate impacts have on the world's most vulnerable people, which include women, children, the poor, and the marginalized. The relationship between the most vulnerable and water is a good starting point for action research on climate change adaptation.

Inherent to this process is the need to recognize that "knowledge is in the room." Facts and knowledge are products not only of academic institutions and research bodies, but also of communities and grassroots movements. In identifying solutions that work best at a local level, traditional knowledge and community-based knowledge are vital.

In order for climate justice work to be participatory and include the most vulnerable, the process must be about more than just representation. At the start, we can look around and ask, "Who is not here?" Is everyone affected by this issue, and by our work, part of our own discussions and decision-making? However, other important questions to ask are about the issues of power and how the relationship between those present brings about change.

Because international agencies stress the importance of representation, most often in the form of gender analysis, project proposals tend to include indicators to measure representation and who participates in training. However, the quality of participation and the impact is often harder to measure and discuss. We can use gender as an example in examining the dynamic between representation and participation.

Women are often identified as those most vulnerable to climate change, so gender is an important lens to consider when examining interventions for climate change adaptation. The United Nations states, "Women are more vulnerable to the effects of climate change than men—primarily as they constitute the majority of the world's poor and are more dependent for their livelihood on natural resources that are threatened by climate change" (UN Women Watch, 2009, p. 1). Extreme weather events and changing climatic patterns affect women's access to and use of water for agricultural, income-generating and household purposes. Furthermore, women are generally more aware than men of the needs of the household—how much water is needed for drinking, cooking, washing, and even sanitation. Women are also aware of needs for water in relation to reproductive and maternal health and caregiving. This responsibility should situate women at the center of discussions about climate change's impacts on household water resources.

The effects of water scarcity on low-income communities compound the gender dimension. Ethne Davey, Chairperson of the Gender and Water Alliance, points out that inadequate access to water is not gender-neutral in its consequences (Davey, 2012). The important role of women in water governance cannot be underestimated.

This said, the concept of gender can be very misleading. It can construct societies in ways that outside funders want to see, or that some participants may find strange. It can ask questions that do not seem important to some participants. Gender equity—having equal numbers of women and men as participants—is not necessarily a meaningful measure of people's contributions to a project. Having good representation from women may not necessarily bring about the best outcomes. For example, at times a man may be able to contribute more meaningfully and think more openly about gender implications than a woman. Or women, in a particular community, may exert power "behind the scenes" which is not readily apparent if you look at who is present at formal meetings. And even if women are present, their voices might effectively be overshadowed by men's voices or even by those of other women who hold different class, ethnic or other social positions.

Building good representation is important. The Kilimanjaro Initiative uses a quota system in its sports programmes to encourage the participation of girls in its programming (e.g., one girl must be playing at all times on their soccer teams, even if most of the players are boys). This approach builds bridges between boys and girls, as boys need to recruit girls to play on their teams and girls become an integral part of the process. However, even with strong female representation, community meetings and processes can still be an unequal playing field.

Examining the barriers of girls and women to being heard once they are present is also important. Do women have equal access to raise their concerns? Do women have equal access to community and broader political structures where collective actions are discussed and decided? Gender awareness training can be a simple way of initiating a discussion with all participants about the different impacts that climate change and changing water resources have on women. These discussions can also be a gateway to talk about other justice dimensions, as well.

In planning our work, we can try to map the participatory challenges for each activity or initiative—since they will differ depending on the situation and circumstances. In evaluating and reflecting on our work, we can consider how participation affected what we did, and the outcomes. We can develop guidelines or questions to ask ourselves about the quality of participation from all those affected.

The dynamics between representation, equality and participation can be discussed in a larger framework of social power. In understanding social power, it is important to consider three main instruments through which power can manifest itself. Paul Speer and Joseph Hughey describe these three instruments using the theories and concepts formed by community organizers such as Saul Alinsky and others.

To begin, Saul Alinsky determined that social power in its most basic understanding determines that those with the greatest resources (financial or social capital) have the greatest power (Speer and Hughey, 2008). A community process, such as the workshops and meetings convened by project partners, may have strong representation from women, community leaders, youth, and the marginalized, but still those with the greatest resources within that particular setting will likely influence the progress or outcome from that group.

Second, social power can also manifest itself in the ability to construct barriers to participation or eliminate barriers to participation through setting agendas and defining issues (Speer and Hughey,

2008). By controlling the topics, timing of discussion, and range of discourse on a topic, those with relatively more power can effectively limit the ability of relatively powerless people to participate and include their perspectives in public debates.

The third instrument of social power concerns the way people think—the shared consciousness that is expressed through myths, ideology, and control of information (Speer and Hughey, 2008). This means that social and cultural norms or even ideas propagated by leaders can determine the outcome of community discussions and decision-making. The response to addressing these forms of power lies in balancing the scale of power with local knowledge and community-based work.

Community-based organizations have acknowledged that these instruments of power are a reality in their work, and highlighted how local knowledge plays a crucial role. In Nairobi, the Kilimanjaro Initiative, which works with youth in Kibera slum, knew that in order to broach the subject of climate adaptation and water governance it would need to do so through a medium which was comfortable to the community members it works with—in this case, football. The use of sports as a focal point led to community-based efforts to manage flooding of a local football pitch in anticipation of heavy rains and swelling of the nearby river. To the youth, the football pitch was an important community resource and a safe place of which they were proud. This nuanced understanding of how local youth perceive and care about their surroundings acted as a gateway to address other climate-related impacts and water issues in the community.

In South Africa, Umphilo waManzi recognized that local communities could help create a history of how water resources have changed in recent years by drawing on the knowledge of residents. They realized they could also create timelines of significant storms or droughts and how local communities adapted in response. This information could be paired with local climate modelling and other research studies to create a balanced knowledge of local climate impacts, from the perspective of local residents. Nakashima *et al.* (2012) argue that "Community-based and local knowledge may offer valuable insights into environmental change due to climate change, and complement broader-scale scientific research with local precision and nuance."

Local governments are at the forefront of addressing the challenge of climate change and urbanization (LOGOTRI, 2011). However, in some cities in sub-Saharan Africa and elsewhere, the priorities and resources of local governments are stretched in addressing service delivery, housing, public health, poverty alleviation, and other pressing urban challenges. In cities where local governments are unable to respond with policies and strategies for climate change adaptation, other stakeholders can play a significant role in helping local governments develop inclusive solutions (LOGOTRI, 2011). Civil society organizations are situated to play this important role. In most cases, this means working within communities to educate children, youth, and adults about climate change impacts and governance, in order to create more informed citizens and as a result, more resilient communities. Civil society groups can help build a base of knowledge and information that can be used in collaboration with governments, researchers and international partners to build responses for climate adaptation and water governance.

In climate justice and water governance work at the community level, participants, organizers, and others involved should create opportunities to reflect on issues of power in their planning and strategizing. Emphasizing local knowledge and experience is also important. Nakashima *et al.* (2012) argue that "Resilience in the face of change is rooted in indigenous knowledge and know-how, diversified resources and livelihoods, social institutions and networks, and cultural values and attitudes." The application of local knowledge can strengthen responses to climatic changes and build resilience across communities, especially when these community-based responses from the neighbourhood level are shared more broadly as well.

13 Durban, climate change, and socio-political transformation

Rehana Dada

Climate change is the most unjust possible situation that human activity could have created. It is effectively a result of the by-products of what has turned out to be an immensely indulgent lifestyle in a handful of countries. It is likely to cause—and we are already seeing some of these impacts—enormous shifts in how life interacts on the planet, in how ecosystems function, and in how human societies feed, water and warm themselves. A disaster on this scale, with the sorts of resource scarcities, impacts on human security, and conflicts that are likely to emerge, should be classified a crime against humanity and all of life. The most vulnerable and the most affected people are the least responsible for the problem. And those most responsible continue to avoid commitments or real action on climate change.

But even if the world entirely stopped emitting carbon tomorrow morning, there would still be impacts from human-caused climate change, and there would still be a need to find ways to cope with those impacts. The sooner and stronger climate change adaptation measures are implemented, the lower the threat that climate change presents, especially in the short term.

There are many actions South Africans can take at national, regional and community level, to ward off the worst, or to adapt to the changes. In many senses, adaptation echoes sustainability practices, with education and understanding critical for ensuring the development of effective adaptation plans that are relevant to unique local circumstances. Adaptation in no way reduces the need for meaningful mitigation on an international scale, but it does offer the potential to significantly reduce the direct impacts on vulnerable peoples. What it does demand, though, is active participation at the community level and the development of plans in full partnership with those who will most benefit from adaptation action.

Encouraging individual, household and community adaptation

Rural African communities know how to cope with a very temperamental climate and over many centuries have developed innovative adaptation mechanisms for natural disasters. This includes knowledge on selecting seed for a variety of weather conditions, and community networks that are able to provide a support buffer for the most affected households during a disaster. However, climate change is expected to increase the frequency of extreme events, and this is likely to reduce community and household ability to cope. In addition, in many cases new types of information may become important, and communities will need to seek solutions outside of their traditional sources of information.

The benefits of adaptation are immediate and usually local, but there are also longer-term positive impacts that are ultimately of benefit nationally and globally. Adaptation actions could, in fact, be beneficial even without the looming threats posed by climate change. Many South African communities already face a complexity of stresses such as land degradation as a result of overpopulation in the ex-homelands (a result of forced resettlements during apartheid), water abstraction for commercial agriculture and timber plantations, and poverty, which combine to increase the vulnerability of these communities to the impacts of climate change.

Adaptation solutions are not often much different from sustainability practices to address existing problems, and it is highly valuable to draw on experiences from sustainability projects to plan and implement climate change adaptation. For example:

- Educating children, and girls in particular, improves people's ability to learn about and cope with changes in their environment.
- Improving maternal health is also likely to reduce the impact of malaria and other diseases.
- Providing clean water and energy will improve health and reduce people's vulnerability to climate change impacts.

Successful adaptation will require involvement and action at individual, household and community level; many of the coping mechanisms will need to be developed locally to be most effective for specific local conditions. In many cases, people will develop adaptation mechanisms organically, as they would deal with any other changes in their environment, and there is some evidence of this occurring already.

Nationally, it will be valuable to develop and implement policy and actions to enhance adaptive capacity of the most vulnerable population groups, such as rural communities in the drier parts of the countries, or the elderly in cities. On a regional level, for example, local governance structures could play a role in protecting ecosystems from development through rezoning and thus enhance the capacity of those ecosystems to attenuate floods and droughts.

Education and awareness are, as always, a crucial starting point for behavioural change. Education/awareness about climate change should focus on developing and implementing actions that people can relate to and see the benefits of immediately.

Climate change in South Africa

In very general terms, South Africa is expected to become warmer throughout the country, drier in most of the country but with increased rainfall to the east of the Drakensberg escarpment.

Most of the country is classified as semi-arid, and already faces severe water pressures. The country's main economic hub, Gauteng province, imports water through interbasin transfers within the country, and also with a significant import from Lesotho. The pressures on the water resources within South Africa include heavy and in many cases unsustainable abstraction for agriculture, timber, mining, urban use and industry. The bulk of the water is used in agriculture, and used in mostly very inefficient irrigation schemes in commercial agriculture.

Pollution from industry and urban settlements is significant, but the key impact on water resources is from mining. Acid mine drainage already threatens water quality significantly in many areas in Gauteng and the North West province, but as mines shut down and there is no longer a pumping of the mine floodwaters, it is expected that the rivers and groundwater resources in these areas will become severely contaminated—bearing in mind that we already see at least one large dam rendered useless for human and agricultural use due to acid mine drainage. In addition, there are numerous new applications for mining and prospecting in Mpumalanga, which if approved could literally poison rivers of importance to the western half of the country, including the Free State and the arid North West province. The prospect of natural gas fracking operations in the Karoo threatens water resources in yet another vast area.

The impacts of climate change on South Africa include significantly increased water stress, significant loss of biodiversity particularly in areas with high endemism or biodiversity hotspots such as the Succulent Karoo and the Cape Floral Kingdom, migration of animal species, increased food security risks, increased poverty, and increased human migration with potential for conflict.

The country has made weak commitments to mitigation despite being responsible for about half of Africa's greenhouse gas emissions, and heavily supports the carbon markets. It has also committed to further significant increases in emissions through its coal energy development, ignoring the implications for air and water pollution. Adaptation planning is progressing in some urban areas but there is little in the public domain about national level adaptation.

Climate change in Durban

Durban has a semi-tropical climate and benefits from being to the east of the Drakensberg escarpment, a key South African "rain trap." It is considerably less pressured in terms of water availability than most of the rest of the country, but there are stresses on the city both in terms of increasing water demands and future water availability.

Predicting climate change impacts is an ongoing challenge. There is no precision and the models are not at high enough resolution to provide detailed predictions, but judging from trends and using the information that is available, it is possible to make some predictions.

Durban is expected to become warmer, with daily maximum temperatures increasing by 2–3 °C and daily minimum temperatures by 3–4 °C over all seasons. An increase in heat waves is expected over the October to March period with temperatures over 30 °C, as well as a higher number of days with temperatures over 30 °C.

In the short term it is not expected that there will be significant changes to rainfall, but by mid century an increase of 10–20 percent is expected, going up to 30–100 percent by 2100. There will also be an increase in the number of days with more than 10 mm of rainfall, but also a significant increase in evaporation. Modelling shows no significant change in the frequency and intensity of droughts. Further detailed analysis will provide a more accurate indication of the situation in the region.

As a result of these changes, expected environmental impacts include biodiversity loss, significant decreases in water availability, increased health challenges, decreased agricultural activity, an increase in storms and flooding events, and ecosystem loss.

Migration to the city from rural areas and other parts of South Africa will increase Durban's population and increase the demands on existing infrastructure and resources. Environmental refugees are already evident, with people moving to the city because of loss of land in rural areas for various reasons, including timber plantations, in addition to widespread stresses such as low livelihoods potential and immigration from other countries. An inability to cope with the needs of migrants could lead to destabilization, displacement of some groups, and conflict.

Water

Increased rainfall is likely to lead to high average runoff and stream flow, potentially providing for replenishment of dams and the ecological reserve. But it is also expected that there will be an increase in evaporation, low absorption of rain into soils, and higher-impact runoff, all of which is likely to lead to increased pressures in some systems. Work is in progress to develop a better understanding of what is likely to happen in terms of water availability in eThekwini (the greater Durban region) as a result of climate change.

Durban's existing population is already straining its water systems, and the population is increasing. There is also the potential additional stress expected in the near future of the city becoming a centre for climate change refugees from rural areas and drier parts of the country.

Higher rainfall is potentially valuable for the city and could create opportunities for household rainwater harvesting, as well as adding to groundwater stores and the ecological reserve. To be able to ensure adequate supply for household and economic use, the water will need to be captured, purified and allocated. The municipality is expecting to face additional challenges with regard to storage and distribution of water. Even with higher rainfall, the increasing demands on the reserve and the periods of particularly dry or wet weather will place pressure and additional strains on health systems.

In rural areas, improved water storage and management could be used to extend water access and support rural livelihoods. Less reliable water availability could exacerbate the existing struggle for good and clean water, leading to increased malnutrition, dehydration and the prevalence of diseases such as cholera, with severe impacts on people living with HIV and AIDS.

Floods

The expected increase in frequency and intensity of flood events as a result of climate change will place additional demands on the municipality's response capacity to storms and flooding which ensures temporary shelter and food, and reconstruction work. Coping with this might need to take on new dimensions such as the need for new medical treatment, longer term and more extensive refugee accommodation, new transport infrastructure, and assistance with new livelihood options.

Floods have impacts on human security and health, and ecosystem health. They disrupt and damage infrastructure including roads, bridges and railway lines, communication and electrical infrastructure. They can also lead to water contamination if water systems are compromised.

Higher rainfall is also likely to expose communities to more frequent localized flooding, especially if stormwater systems are unable to cope with the frequency or intensity of floods. In areas or communities where clean water is not provided, higher rainfall could lead to greater incidences of water contamination and associated health complications. Flooding also offers opportunities for new breeding grounds for disease vectors.

Communities and properties in low-lying areas would be inundated and this could lead to displacement during emergency periods, or for the long term in some cases. Shack settlements are especially vulnerable because of poor construction, often in areas prone to flooding, with no stormwater systems, and high concentrations of people.

Rural communities could become isolated, preventing support from emergency services and increasing the risk of disease. There are also psychological concerns such as the effects of trauma and depression following loss of friends and family, homes and livelihoods. These personal losses increase stress and hamper people's ability to recover and return to a normal life.

Health

Vulnerable populations such as the elderly and people living with HIV and AIDS are likely to become more pressured as a result of heat stress, food insecurity, and exposure to disease outbreaks.

Increased heat stress will affect people across all income groups, but especially more vulnerable groups such as children and the elderly, and people with poor access to clean drinking water.

Cholera and other water-borne diseases are likely to become more prevalent as a result of floods.

Small changes in temperature and precipitation can lead to malaria outbreaks, a disease which is likely to increase its presence in Durban. The municipality's existing malaria program will need to be upgraded to cope with the increased stresses expected.

The sea

Natural systems in Durban can cope with a sea level rise of up to 20 cm, but it is expected that by 2100 sea level rise could reach 50 to 90 cm. Durban is already experiencing an increase in sea levels, at a rate of 2.7 mm/year between 1970–2003. This is expected to continue and accelerate.

Sea level rise is projected to cost R134 billion (USD$16.3 billion) if the annual direct and indirect costs are considered, such as the impacts of coastal erosion on tourism and inundation on property and industry.

Durban is hilly, so most of the city will not be inundated, but several sites are at risk. This includes industrial areas in the Durban South Basin, a number of transport hubs including the harbour, beaches that are important for tourism, parts of the city centre, and some agriculturally and ecologically important areas to the south of the city.

Higher sea levels and increased erosion will impact coastal ecosystems such as mangroves, estuaries and coral reefs, which are important breeding and feeding grounds for many economically and ecologically important marine fish species. This will affect fish stocks and biodiversity, with impacts on livelihoods and tourism. There may also be salination of some aquifers which would reduce freshwater availability. Changes in rainfall that affect coastal ecosystems reduce the ability of these ecosystems to act as buffers for the impacts of sea level rise and sea storms.

It is likely that Durban will be impacted by Indian Ocean cyclones or the effects of cyclones moving further south, with more severe and more frequent storms. This will lead to increased coastal erosion, and damage to coastal ecosystems and infrastructure such as roads, railway lines, bridges and homes.

Farming

eThekwini Municipality, of which Durban is the core, includes large areas of agricultural land including many subsistence farms that are rain-fed. Changes to rainfall distribution and seasonality will place increasing strain on farming, and subsistence farming in particular, because of crop and livestock losses. This will affect local food security.

If subsistence or small-scale farms that rely on rain do not have the profit margins to afford irrigation, or if there is inadequate water for irrigation, this could result in increased poverty and reduced food security.

Temperature rise is likely to result in a shift of agriculture to higher and cooler areas, changes in the types of crops planted, and an increase in the need for irrigation. For the period 2070–2100 it is already expected that the irrigation demands on Midmar Dam, the main supply for Durban, will increase by 25 percent. Moreover, during this same period, the Umgeni catchment, which is the main catchment for Midmar, is expected to have a reduction of water availability by an amount of water equivalent to that stored behind the dam.

Existing pressures on the water and land resources that affect agricultural productivity include degradation from bad land use practices, proliferation of alien invasive plant species, timber plantations that increase erosion and reduce biodiversity, and water extraction for timber and agriculture. Timber is also expanding into communal lands through so-called "woodlot"' schemes which are considered to be an extension of commercial plantations and ultimately benefit only timber companies.

Biodiversity

Durban is an unusual city in that, despite its population concentration, it has large natural areas, mainly protected as nature reserves or through the Durban Metropolitan Open Space System (DMOSS). There are still areas of land near the urban centre that are undeveloped, and strips of land in between developments that are left untouched. There are impacts on these areas, including invasive alien species, but nature thrives. For example, vervet monkeys and mongoose are fairly common residential visitors, as are a fantastic range of birds and reptiles. The area is home to a wealth of diversity and includes coastal dune forests, grassland, scarp forest and wetlands. It is home to 73 threatened species.

Biodiversity is sensitive to changes in temperature and rainfall, and although there are no clear data for Durban specifically it is expected that by 2050 South Africa will see a reduction of 50 percent in areas covered by current biomes. In other parts of the country, shifts in species migrations are already evident. There is expected to be an increase of invasive alien species which will impact on biodiversity, ecosystem stability, economic activities, erosion, and water availability.

Increased sea surface temperature is likely to mean an increase in the migration of species living along the coast to higher altitude eastern escarpments. It is expected that there will be large extinctions, and that most animal species will become concentrated in higher eastern escarpment regions by 2070–2100. Sea level rise and storm surges are likely to affect dune vegetation and mangroves and floods are likely to impact on riparian and other vegetation, with roll-over impacts on fauna.

Durban's municipal adaptation priorities are outlined in Box 13.1.

Box 13.1 eThekwini Municipality's adaptation planning

Durban's municipal adaptation focus includes the following, as stated in a municipal document intended for the public (Lewis, 2011).

Disaster Risk Reduction

- Improve early detection systems for storms and other natural disasters.
- Enhance early warning communication to public.
- Develop community disaster management plans and ensure awareness of plans in vulnerable locations.
- Establish effective post disaster recovery and support plans across all sectors.
- Assess and improve resistance of infrastructure to extreme weather events.

Health

- Expand existing malaria prevention programme.
- Secure sustainable energy and clean water sources for health provision.
- Identify vulnerable groups—children, elderly, poor—and improve surveillance of disease outbreaks.
- Heat emergency plans.
- Education campaigns about heat stress, environmental health associated with climate change.

Food security

- Promote local food production within the municipality through small-scale farmers' production.
- Educate farmers on drought resistant crops, erosion prevention, efficient water use practices.
- Establish community food reserves for emergency situations.

Water and sanitation

- Increase use of water recycling and promote water storage in cisterns and rainwater catchment tanks.
- Reduce water losses from municipal water systems—better leak detection and flow and pressure controls.
- Reduce water demand from sanitation systems by using dry sanitation systems such as urinary diversion toilets and

developing systems that use grey water for toilet flushing, low flush toilets, etc.

- Initiate water conservation education campaigns and community water development assessments.
- Promote water conservation and possibly water trading for industries.
- Assess the effects that increased heat, storms and sea level rise will have on water provision and sanitation infrastructure, and upgrade or relocate where necessary.

Coastal, stormwater and catchment management

- Reassess areas vulnerable to flooding using predicted future rainfall changes, notify property owners, and restrict proposed developments in these areas.
- Prevent development in areas predicted to be affected by "one in fifty-year" storms, and coastal areas which will be underwater due to sea level rise over the next 50 years, by adjusting setback lines to climate change scenarios.
- Develop flood prevention and emergency response plans for areas that will become more prone to flooding.
- Insist that new developments have stormwater management plans and flood attenuation features.
- Incentivize and assist those in the most flood prone areas to relocate.

Infrastructure

- Ensure that new or upgrade infrastructure are not located in flood prone, low-lying or coastal areas, and materials and designs account for increased temperature and severe weather.
- Reduce energy demand through energy efficiency and on-site renewable energy production.
- Encourage use of public transport.

Strategic planning and economic development

- Revise spatial development framework to prevent new developments in potentially hazardous areas by rezoning them.

- Assess vulnerability of key industries to climate change and promote efficiency in these sectors.
- Establish a coordinating mechanism to ensure that climate change predictions are accounted for in all sectors of municipal planning.

Biodiversity

- Assess the vulnerability of indigenous habitats and species.
- Revise DMOSS to ensure protection of refuge areas for climate sensitive species and maintain corridors that will allow these species to migrate to refuge areas as the temperatures and rainfall change.
- Increase removal of alien invasive species, especially those affecting water supplies and those adapted to warmer and drier conditions.
- Protect and rehabilitate ecosystems such as wetlands and mangroves that play a role in buffering storms as well as natural areas such as coastal forests that play a role in reducing heat in the city, filtering and absorbing water, and buffering floods.

What does adaptation offer?

In a very real sense, adapting to climate change demands a strong rethink of current development practices, and almost forces sustainability options to the forefront. For example: biodiversity protection requires the safeguarding of natural areas, creation of new nature reserves and nature corridors; food security means ensuring that resources such as soil and water remain healthy; increasing the resilience of vulnerable communities includes education, human rights, health, and strong participation and co-planning; and ensuring economic growth involves intelligent management of water resources and other natural resources, as well as education and the development of well informed work teams. Box 13.2 outlines some ideas for use in planning community climate change education and action workshops.

In order to ensure that as many people as possible weather the climate change storm, people must collectively build a more intricate and respectful relationship between human development, nature and economic development.

Box 13.2 **Starting point: questions for community education and action on climate change**

What do we need in order to build our own capacity to cope with climate change?

1. Understanding climate change.
    a. What is climate change?
    b. How will climate change affect my home, community, farm or business?
    c. Can climate change be stopped?
    d. What are my local authority and national government doing to stop further climate change?
    e. What can I do to stop further climate change?
2. Planning for climate change.
    a. What can I do to protect myself and my livelihood?
    b. What can I do to make sure that I have access to food, water and energy now and in future years?
    c. How do I make sure that I and my loved ones are informed about climate change and adaptation?
    d. What do I already know and what do I already have that I can use to help me cope with climate change?
    e. What do I need in addition to this?
3. Government and institutional support.
    a. Who in my community can help me find the information and resources that I need?
    b. Who in my community can help me develop an adaptation plan?
    c. What help can I get from: Agricultural extension officers / Health workers / Teachers / Experienced community members with knowledge about disasters / My community leaders?

4. Sharing experiences.
    a. How can I help to make sure that as many people as possible in my own community know as much as possible about how to adapt to climate change?
    b. How can I learn from other communities across the world who are working on adaptation for climate change?
    c. How can I share what I have learnt with neighbouring communities and people elsewhere in my country and across the world?
5. Assessing and learning.
    a. What have I done that's worked?
    b. How can I improve the concepts or projects that worked less well?
6. How can I get my government to improve its national scale planning for adaptation on a national scale?

14 Conclusion

Developing community-based responses to climate change

Patricia E. Perkins and Patricia Figueiredo Walker

The stories in this book are specific examples of climate change resilience-building from the grassroots which are taking place worldwide, driven by the immediacy and urgency of climate change and extreme weather events. A range of additional examples is outlined in Box 14.1. More examples are emerging every day.

Box 14.1 **Additional examples of grassroots climate change initiatives**

- The grassroots non-governmental organization (NGO) Agua Doce organizes flood protection, socio-environmental education, conservation and local job creation for community development in the Guanabara Bay area near Rio de Janeiro, Brazil (Leary and Perkins, 2011).
- The Ecoar Institute for Sustainability carries out socio-environmental education, green jobs, and recycling projects in São Paulo, Brazil (http://www.ecoar.org.br/web/pag.php?id=31).
- The Centre for Socio-Environmental Knowledge and Care of the Paraná Basin (Centro de Saberes) leads watershed-based education and community development programmes such as "Cultivating Good Water" (Cultivando Agua Boa), building resilience in the Paraná/Plate basin in southern Brazil, Paraguay, western Bolivia, and northern Uruguay and Argentina (http://www.cultivandoaguaboa.com.br/o-

programa/centro-de-saberes-e-cuidados-socioambientais-da-bacia-do-prata).

- The Riberas Rioplatenses project in Uruguay and Argentina assembles local knowledge on climate change impacts and adaptation priorities (http://www.iied-al.org.ar/riberas).
- Dozens of communities in Arctic Canada share health-related climate change education, arts and community-based initiatives (McClymont Peace and Myers, 2012, http://www.circumpolarhealthjournal.net/index.php/ijch/article/view/18412/html).
- The Green Change Project focuses on green community development and training local residents for green jobs in low-income areas of Toronto, Canada (http://www.toronto.ca/livegreen/pdf/jane_finch.pdf; http://yorkwestadvocate.wordpress.com/2010/05/13/jane-finchs-green-change-project-wins-toronto-green-award/).
- The Indian Youth Climate Network organizes local climate change action projects and develops youth leadership on climate change (http://www.iycn.in/content/about-iycn).
- Majora Carter leads green job creation and climate justice projects in the South Bronx, USA and elsewhere (http://itsgettinghotinhere.org/2007/11/01/majora-carter-says-showing-up-for-climate-solutions-is-more-than-half-the-battle/; http://www.theroot.com/views/greening-ghetto).
- Indigo Development and Change carries out local conservation, leadership development, environmental education, and green job creation/climate justice initiatives in Western Cape province, South Africa (http://www.indigo-dc.org/).
- In the area devastated by Hurricane Katrina in New Orleans, many organizations carry out erosion control, environmental testing and awareness-building, and environmental education, such as Bayou Rebirth (http://www.bayourebirth.org), Coast Currents (http://www.coastcurrents.org/), LA Bucket Brigade (http://www.labucketbrigade.org), and the Gulf Restoration Network (http://healthygulf.org/).

"Bottom-up," or community-based approaches to climate change adaptation generally make use of local knowledge to identify a community or group's vulnerability to current and future climatic hazards and to create coping strategies and mechanisms which are tailored to that community or group's specific needs. These approaches seek to lessen vulnerability to climate change by strengthening resilience—through training, capacity-building, and literacy.

As Ensor and Berger have noted:

> (adaptation) activities should simultaneously reduce the impacts of potential climate change and improve the well-being of households or communities in the short term—for example through addressing poor housing, degraded soils or the inequitable distribution of resources. These approaches to adaptation are in essence vulnerability reduction measures, and can be classified as no-regrets strategies—meeting climate change adaptation goals while fulfilling broader development ends.
>
> (Ensor and Berger, 2009: 16)

If improperly or not at all addressed, climate change can exacerbate gender and structural inequalities, poverty, and injustice, hence the need to create policies and adaptation strategies which not only address the negative impacts of climate change, but improve the well-being and living conditions of the poor and vulnerable.

Furthermore, community-based approaches to climate change adaptation foster and strengthen political participation by vulnerable and marginalized people, as they often link community, government and civil society organizations.

> Past history shows that the poor and vulnerable do not get a fair share of resources unless they can mobilize effectively and there is parallel pressure on the powerful to make decisions in favour of the many, not the few. In the context of the climate change negotiations, this means making sure that the voices of ordinary Africans—women, men, young, old, farmers and slum dwellers—are heard loud and clear as the policies and institutions for addressing the most challenging of global problems are developed.
>
> (Toulmin, 2010: 152)

As mentioned throughout this book, women in the Global South are particularly vulnerable to climate change due to their gendered roles—collecting firewood, processing food crops, caring for the young and elderly, and other household tasks. Because women play significant roles in their communities, building their adaptive capacity is likely to improve the adaptive capacity of fellow community members, therefore strengthening the overall resilience of the community. As Chapter 4 indicates, involving youth in climate change adaptation initiatives can yield similar results.

Creating the conditions for women, youth and all who are marginalized, or otherwise discriminated against, to become active agents of change within their communities, municipalities, and beyond helps to drive the climate justice movement forward, mitigate the impacts of climate change and create stronger, more resilient individuals and communities. Some practical ways of doing community-based environmental education and climate change organizing are outlined in Box 14.2.

***Box 14.2* Ways of doing community education and political engagement on water and climate change**

1. Education initiatives
    a. Watershed learning circles: Watershed committees, assisted by civil society organizations (CSOs), bring together key civil society groups from throughout the watershed (e.g. journalists, teachers, artists, government officials) to discuss and develop needed action programs on water-related issues.
    b. Leadership training: CSOs work with government and watershed committee officials to develop inclusivity training and confidence-building workshops for special groups such as women, youth, or seniors, to welcome and facilitate their participation in water governance institutions.
    c. Gender/equity awareness workshops: Community organizations facilitate information-sharing on how and why different groups are affected by climate change and have differing water access (women, ethnic groups, seniors, etc.).

2. Public outreach initiatives

   a. Water dialogues: local residents discuss specific water issues with government officials, in forums facilitated by civil society organizations.
   b. Learning journeys: community groups run tours of key water/environment-related features in the area for local residents, government officials and visitors.
   c. Water conflict mediation training: special NGO-run training programs address water conflicts (e.g. over standpipe access, fugitive emissions of pollutants, riparian rights, etc.) through skills training and information on how to access government supports.

3. Creative and cultural initiatives

   a. Photo-voice: community members photograph local scenes significant to them, in relation to climate change effects, and discuss/share with government officials.
   b. Community mapping: residents collectively draw maps of important water features in the neighbourhood and how extreme weather affects them; then discuss.
   c. Water walks/storytelling parades: using waterways as a focus, local residents and visitors tour stream beds and floodplains, stopping for special presentations on local history, wildlife, food production, music, community assets, etc.
   d. Collective storytelling workshops: community groups use selected "props" to spark creativity as they compose a joint story about the local watershed, sharing personal memories.
   e. Sports field and league development: CSOs and youth groups organize social/recreational activities to make use of floodplains and advocate for their preservation as open, public space.

4. Conservation initiatives

   a. Community-based water monitoring: community groups work with government authorities to monitor pollution, biostatus, and flooding/drought in local waterways, and share the results.

b. Watershed restoration: CSOs organize cleanups of local waterways to improve local agriculture, fishing, and recreation opportunities as well as environmental awareness.

5. Subsistence initiatives

a. Water harvesting: CSOs work with household members to develop ways of retaining rainfall from roofs and yards for home and garden use, and spread related practices.
b. Community gardening: CSOs locate space and train and organize community members to plant gardens for collective food production.
c. Community kitchens: NGOs or churches establish kitchen space for collective food preservation, processing and cooking and for feeding vulnerable community members and buffering time pressures for women.
d. Dry sanitation: CSOs facilitate the distribution and education/advocacy/promotion of urine diversion toilets.

6. Job-creation initiatives

a. Green change agents training: NGOs run workshops for local youth in certificate programs helping develop job-related skills such as environmental building audits, green construction, green roof landscaping, floodplain rehabilitation, etc. The NGOs also set up internships and job placement opportunities for participants.

7. Networking initiatives

a. Academic-activist linkage building: students get academic credit for internships with local civil society organizations, and help document their work and write funding applications; professors research and disseminate the methods and accomplishments of community groups.
b. Watershed networking among universities, grassroots organizations, and/or government officials: through

participatory research and knowledge-sharing, individuals from different groups build personal relationships and trust as well as sharing information.

Adapted from P. Figueiredo and P. E. Perkins, (2012) 'Women and Water Management in Times of Climate Change,' *Journal of Cleaner Production.*

Such climate justice initiatives generate innovative and specifically-tailored ways of strengthening the ability of civil society to participate in watershed decision-making, within the institutional structures set out by national water laws and other statutes.

Academic and international partnerships can strongly contribute to these locally-grounded initiatives. For example, graduate students from partner universities can visit each other's campuses for short research trips; faculty members can collaborate and orient the research, while recognizing that integral collaboration of community-based NGOs is essential. Students can help civil society organizations to document and capitalize politically on their work at the grassroots. CSOs, with their long experience on environmental education and organizing in low-income communities, are often particularly good at presenting information clearly to politicians and bureaucrats. There are many examples of such synergies between research and development-oriented activism.

The researchers and activists working together on such projects will share a change-oriented approach to exploration and communication: What water-related issues are the most important to local communities? What are the best, most creative and interesting ways to help get the public involved in and knowledgeable about water issues? When the law creates space at the decision-making table for "civil society," how can that democratic opening be used effectively and how can multiple publics and views be included? These questions are relevant worldwide, in the global North as well as the global South, and in both urban and rural areas.

Climate justice clearly requires new kinds of governance, which recognize equitable access and shared responsibilities for managing common resources: water, air, and ecosystems which we all rely on in interrelated ways for our livelihoods.

As Nobel prize-winning economist Elinor Ostrom noted in her book *Governing the Commons*, it is at the local level that people can, under the right circumstances, generate the trust necessary to set up governance institutions which are complex and flexible enough to manage common property resources sustainably.

> . . . (I)n the smaller-scale common property resources (CPRs). . . . individuals repeatedly communicate and interact with one another in a localized physical setting. Thus, it is possible that they can learn whom to trust, what effects their actions will have on each other and on the CPR, and how to organize themselves to gain benefits and avoid harm. When individuals have lived in such situations for a substantial time and have developed shared norms and patterns of reciprocity, they possess social capital with which they can build institutional arrangements for resolving CPR dilemmas.
>
> (Ostrom, 1990: 183–184)

The networks, linkages, and interrelationships which are necessary for forging these new governance systems are built when people work together to respond to climate change as they experience it, in each local area and community. Collaborative grassroots climate justice organizing thus establishes the framework for a more sustainable future.

Bibliography

ActionAid (2006) *Climate change, urban flooding and the rights of the urban poor in Africa: Key findings from six African cities*. A report by ActionAid, London and Johannesburg.

Adger, W. N. (2003) 'Social aspects of adaptive capacity,' in Smith, J. B., Klein, R. J. T. and Huq, S. (eds.) *Climate Change, Adaptive Capacity and Development*. London: Imperial College Press.

Adger, W. N. and Kelly, P. M. (1999) 'Social vulnerability to climate change and the architecture of entitlements,' *Mitigation and Adaptation Strategies for Global Change*, vol. 4, pp. 253–266.

Adger, W. N., Arnell, N. W. and Tompkins, E. L. (2005) 'Successful adaptation to climate change across scales,' *Global Environmental Change*, vol. 15, pp. 77–86.

The Agenda (2012) *The Politics of Water* [Podcast]. South Africa: Agenda FM.

Anisfeld, S. C. (2010) *Water Resources*, Washington, DC: Island Press.

Arnall, A. H. (2012) 'Compreendendo a capacidade adaptativa ao nível local em Moçambique,' *Aliança Africana para a Resiliencia ás Mudanças Climáticas (ACCRA)*. Relatório síntese de Moçambique. [Electronic]. Available: http://www.cebem.org/cmsfiles/publicaciones/Comprendiendo_capacidade_moz.pdf (15 August 2012).

Arnell, N. W. (1999) 'Climate change and global water resources,' *Global Environmental Change*, vol. 9, pp. S31–S49.

Arnstein, S. R. (1969) 'A Ladder of Citizen Participation,' *Journal of the American Institute of Planners*, vol. 35, no. 4, July, pp. 216–224.

Asian Communities for Reproductive Justice (2009) 'Looking Both Ways: Women's Lives at the Crossroads of Reproductive Justice and Climate Justice,' Oakland, California. Available: http://www.reproductivejustice.org/ACRJ_Looking_Both_Ways.pdf (13 November 2011).

Bakker, K. (2007) 'The "commons" versus the "commodity": Alter-globalisation, anti-privatisation and the human right to water in the global South,' *Antipode*, vol. 39, no. 3, pp. 430–455.

Bied-Charreton, M. (2008) *Integrating the Combat Against Desertification and Land Degradation into Negotiations on Climate Change: A Winning Strategy*. UNCCD. Available: http://www.unccd.int/science/docs/non_paper_desertif_Climate_eng.pdf (5 May 2011).

Blackden, C. M. and Wodon Q. (eds.) (2006) *Gender, Time Use, and Poverty in Sub-Saharan Africa*, World Bank Working Paper No. 73. Washington, DC: The World Bank.

Boko, M., Niang, I., Nyong, A., Vogel, C., Githeko, A., Medany, M., Osman-Elasha, B., Tabo, R. and Yanda, P. (2007) 'Africa,' pp. 433–467 in M. L. Parry, O. F. Canziani, J. P. Palutikof, P. J. van der Linden and C. E. Hanson (eds.) *Climate Change 2007: Impacts, Adaptation and Vulnerability. Contribution of Working Group II to the Fourth Assessment Report of the Intergovernmental Panel on Climate Change*, Cambridge, UK and New York: Cambridge University Press.

Bond, P. (2002) *Unsustainable South Africa*, London: Merlin Press.

Bond, P. and Dorsey, M. K. (2010) 'Anatomies of environmental knowledge and resistance: Diverse climate justice movements and waning eco-neoliberalism,' *Journal of Australian Political Economy*, vol. 66, pp. 286–314.

Brennan, T. (2003) *Globalization and its Terrors: Daily Life in the West*, New York: Routledge.

Brooks, N. (2003) *Vulnerability, risk and adaptation: a conceptual framework*, Working Paper No. 38, Tyndall Centre for Climate Change Research, University of East Anglia, Norwich.

Brooks, N., Adger, W. N. and Kelly, P. M. (2005) 'The determinants of vulnerability and adaptive capacity at the national level and the implications for adaptation,' *Global Environmental Change*, vol. 15, pp. 151–163.

Bullard, N. (2009) 'The State of Climate Politics,' speech delivered at a Rosa Luxemburg Foundation conference in Johannesburg, South Africa, November 19, 2009.

CIDA (Canadian International Development Agency) (2002) *Gender Equality and Climate Change: Why Consider Gender Equality When Taking action on climate change?* Ottawa: Canadian International Development Agency. Available: http://www.acdi-cida.gc.ca/INET/IMAGES.NSF/vLUImages/Climate%20change3/$file/Gender-2.pdf (6 October 2011).

Davey, E. (2012) *The Politics of Water, Empowering Women for Gender Equity* [Podcast]. Available: http://www.blubrry.com/agenda/1148783/the-politics-of-water/ (28 September 2012).

De Lucia, V. (2009) 'Hegemony and climate justice: a critical analysis' in Böhm, S. and Dabhi, S. (eds.) *Upsetting the Offset: The Political Economy of Carbon Markets*, London: Mayfly. Available: http://mayflybooks.org/?page_id=194 (30 June 2011).

De Marchi, B., Funtowicz S. O., Lo Cascio S. and Munda G. (2000) 'Combining participative and institutional approaches with multicriteria evaluation. An empirical study for water issues in Troina, Sicily,' *Ecological Economics*, vol. 34, pp. 267–282.

Department of Water Affairs and Environment (2009) *Green Drop Report 2009 Version 1*, South African Waste Water Quality Management Performance. Pretoria, South Africa: DWA.

Department of Water Affairs and Environment (2010) *Blue Drop Report 2010*, South African Drinking Water Quality Management Performance. Pretoria, South Africa: DWA.

Department of Water Affairs and Environment (2011) *Blue and Green Drop 2011*, The Quality of your drinking Water and Wastewater Service is our Concern. Pretoria, South Africa: DWA.

Department of Water Affairs and Forestry, Department of Environmental Affairs and Tourism, Water Research Commission, Umgeni Water, eThekwini Municipality, CSIR Environmentek and University of Natal (2002) *State of Rivers Report—Umngeni River and Neighbouring Rivers and Streams.* Available: http://www.dwaf.gov.za/iwqs/rhp/state_of_rivers/state_of_umngeni_02/umngeni_02.pdf (12 October 2012).

De Sherbinin, A., Schiller, A. and Pulsipher, A. (2007) 'The vulnerability of global cities to climate hazards,' *Environment and Urbanization*, vol. 19, pp.39–64. DOI: 10.1177/0956247807076725.

De Villiers, S. (2005) 'The hydrochemistry of rivers in KwaZulu-Natal,' *Water SA*, vol. 31, no. 2, pp. 193–198.

Dessai, S., Adger, W. N., Hulme, M., Turnpenny, J. Kohler, J. and Warren, R. (2004) 'Defining and experiencing dangerous climate change,' *Climatic Change*, vol. 64, pp. 11–25.

Dickens, C. W. S. and Graham, P. M. (2002) 'The South African Scoring System Version 5 Rapid Bioassessment Method for Rivers,' *African Journal of Aquatic Science*, vol. 27, pp. 1–10.

Dyer, R. (1999) *Proceedings of the 9th International Rainwater Catchment Systems Conference. Promotion of rainwater catchment in Southern Africa.* [Online] Available: http://www.abcmac.org.br/files/simposio/2simp_robdyer_promotionrainwater.pdf (28 September 2012).

Enright, W. Interviewed by Mary Galvin (6 July 2010).

Ensor, J. and Berger R. (eds.) (2009) *Understanding Climate Change Adaptation: Lessons from Community-Based Approaches.* Bourton on Dunsmore, UK: Practical Action Publishing.

eThekwini Municipality (2006) *Climate Future for Durban: Phase 2-Headline Climate Change Adaptation Strategy*, CSIR: Pretoria.

eThekwini Municipality (2009) *Climate Change Municipal Adaptation Plan Health and Water*, eThekwini Municipality: Durban.

eThekwini Municipality (2010a) *Guideline for Designing Green Roof Habitats*, eThekwini Municipality: Durban.

eThekwini Municipality (2010b) *Energy Efficiency Guideline*, eThekwini Municipality: Durban.

eThekwini Municipality (2010c) *Sustainable Waste Management Guideline*, eThekwini Municipality: Durban.

Fairbanks, D. H. K. and Benn, G. A. (2000) 'Identifying regional landscapes for conservation planning: a case study from KwaZulu-Natal,' *South Africa Landscape and Urban Planning*, vol. 50, pp. 237–257.

Faucheux, S. and Hue, C. (2001) 'From irreversibility to participation: towards a participatory foresight for the governance of collective environmental risks,' *Journal of Hazardous Materials*, vol. 86, pp. 223–243.

Figueiredo, P. and Perkins P. E. (2012) 'Women and water management in times of climate change: participatory and inclusive processes,' *Journal of Cleaner Production*, DOI: 10.1016/j.jclepro.2012.02.025.

Folke, C., Carpenter, S., Emqvist, T., Gunderson, L., Holling, C. S. and Walker, B. (2002) 'Resilience and sustainable development: building adaptive capacity in a world of transformations,' *Ambio*, vol. 31, pp. 437–440.

Friedman, S. (2005) *Sending Them a Message: Culture, tax collection, and governance in South Africa*, Johannesburg: Centre for Policy Studies.

Galvin, M. (2009) *Straight talk to Strengthen Delivery in the Water Service Sector*. The Water Dialogues, Durban.

Galvin, M. (2010a) 'Unintended consequences: Development interventions and socio-political change in rural South Africa,' in B. Freund and H. Witt (eds.) *Development Dilemmas in Post-Apartheid South Africa*, Durban: University of KwaZulu Natal Press.

Galvin, M. (2010b) 'Survival or advocacy in rural South Africa?' in D. Maharaj and P. Bond (eds.) *Zuma's Own Goal: Losing South Africa's 'War on Poverty,'* Trenton, New Jersey: Africa World Press.

Galvin, M. (2011) 'Participating in urban myths about women's rural water struggles,' Agenda 88.

Galvin, M. (forthcoming) 'Leaving boxes behind: civil society and water and sanitation struggles in Durban, South Africa,' submitted in 2012, to appear in a special issue of *Antipode*.

Graham, P. M., Dickens C. W. S. and Taylor, R. J. (2004) 'MiniSASS—A novel technique for community participation in river health monitoring and management,' *African Journal of Aquatic Science*, vol. 29, no. 1, pp. 25–35.

Government of Kenya (2008) *Draft Nairobi Metropolitan Area Bill*, Ministry of Nairobi Metropolitan Development.

Government of Kenya (2011) *Development of a Spatial Planning Concept for Nairobi Metropolitan Region*. Draft Plan. Ministry of Nairobi Metropolitan Development.

Gregory, R. and Slovic, P. (1997) 'A constructive approach to environmental valuation,' *Ecological Economics*, vol. 21, pp. 175–181.

Grenfell, M. C., Ellery, W. and Grenfell, S. E. (2008) 'Tributary valley impoundment by truck river floodplain development: a case study from the KwaZulu-Natal Drakensberg foothills, eastern South Africa,' *Earth Surface Processes and Landforms*, vol. 33, pp. 2029–2044.

Hamdy, A., Abu-Zeid, M. and Lacirignola, C. (1998) 'Institutional capacity-building for water sector development,' *Water International*, vol. 23, pp. 126–133.

Handmer, J. W., Dovers, S. and Downing, T. E. (1999) 'Societal vulnerability to climate change and variability,' *Mitigation and Adaptation Strategies for Global Change*, vol. 4, pp. 267–281.

Hinchcliffe, F. *et al.* (1999) *Fertile Ground: The Impacts of Participatory Watershed Management*. London: Earthscan/IT Publications.

Holland, A. (1997) 'The foundations of environmental decision-making,' *International Journal of Environment and Pollution*, vol. 7, no. 4, pp. 483–495.

IPCC (Intergovernmental Panel on Climate Change) (2007) *Climate Change 2007: Synthesis Report*. Available: http://www.ipcc.ch/pdf/assessment-report/ar4/wg2/ar4-wg2-chapter9.pdf (4 May 2011).

IRIN (Humanitarian News and Analysis) (2008) 'Mozambique: "The issue of natural disasters is very worrying"', UN Office for the Coordination of Humanitarian Affairs, October 20. Available: http://www.irinnews.org/Report/81004/MOZAMBIQUE-The-issue-of-natural-disasters-is-very-worrying (15 August 2012).

Jaggernath, J. (2011) 'Environmental conflicts in the South Durban Basin: Integrating residents' perceptions and concerns resulting from air pollution,' *African Journal on Conflict Resolution*. Available: http://www.ajol.info/index.php/ajcr/article/viewFile/63316/51199 (13 April 2012).

Kangalawe, R., Mwakalila, S. and Masolwa, P. (2011) 'Climate Change Impacts, Local Knowledge and Coping Strategies in the Great Ruaha River Catchment Area, Tanzania,' *Natural Resources*, pp. 212–223.

Kapoor, I. (2008) *The Post-Colonial Politics of Development*. London/New York: Routledge.

Kasperson, J. X. and Kasperson, R. E. (2001) *Climate Change, Vulnerability and Social Justice*. Stockholm: Stockholm Environment Institute.

Kates, R. W. (2000) 'Cautionary tales: adaptation and the global poor,' *Climatic Change*, vol. 45, pp. 5–17.

Kelly, P. M. and Adger, W. N. (2000) 'Theory and practice in assessing vulnerability to climate change and facilitating adaptation,' *Climate Change*, vol. 47, pp. 325–352.

Kindon, S., Pain, R. and Kesby, M. (2008) 'Participatory Action Research,' *International Encyclopedia of Human Geography*. Amsterdam/London: Elsevier, pp. 90–95.

Kithlia, J. (2011) 'Climate change risk responses in East African cities, needs, barriers and opportunities,' *Current Opinion in Environmental Sustainability*, vol. 3, no. 3, pp. 176–180.

Kothari, U. (2001) 'Power, knowledge and social control in participatory development,' in B. Cooke, and U. Kothari (eds.) *Participation: The New Tyranny?* London: Zed Books.

Leary, A. L. T. and Perkins, P. E. (2011) 'International synergies to address climate change: participatory community organizing in Toronto and the Baixada Fluminense, Brazil,' *The Global Studies Journal*, vol. 4, pp. 23–34.

Lewis, M. (2011) *Climate change adaptation planning for a resilient city*, eThekwini Municipality, 2010/11. Available: http://www.durban.gov.za/durban/services/development-planning-and-management/epcpd/documents/Durbans%20Municipal%20Climate%20Protection%20Programme_Climate%20Chan.pdf (30 June 2011).

Lim, B. and Spanger-Siegfried, E. (2004) *Adaptation Policy Frameworks for Climate Change: Developing Strategies, Policies and Measures*. Cambridge, UK: Cambridge University Press and United Nations Development Programme.

Lin, J. and Biyela, P. T. (2005) 'Convergent acquisition of antibiotic resistance determinants amongst the Enterobacteriaceae isolates of the Mhlathuze River, KwaZulu-Natal (RSA),' *Water SA*, vol. 31, no. 2, pp. 257–260.

Lin, J., Biyela, P. T., Puckree, T. and Bezuidenhout, C. C. (2004) 'A study of the water quality of the Mhlathuze River, KwaZulu-Natal (RSA): Microbial and physico-chemical factors,' *Water SA*, vol. 30, no. 1, pp. 17–22.

LOGOTRI (Network of Local Government Training and Research Institutes in Asia and the Pacific) (2011) 'Climate Change Resilience for Sustainable Cities,' *Cities of Opportunity: Partnerships for an Inclusive and Sustainable Urban Future*. Available: http://www.unescap.org/apuf-5/bazzar/summary-of-discussion/1-Climate-Change.pdf (13 September 2012).

Love, D., Magombeyi, M., Munguambe, P., Mupangwa, W. and Ncube, B. (2008) *Methodologies and case studies for investigating upstream-downstream interactions of rainwater water harvesting in the Limpopo*

Basin. Available: http://www.waternetonline.ihe.nl/challengeprogram/P68%20Ncube%20RWH.pdf (28 September 2012).

Luxemburg, R. (1968) *The Accumulation of Capital*, New York: Monthly Review Press.

Macleod, N. Interviewed by Mary Galvin (30 April 2010).

Macleod, N. Interviewed by Mary Galvin (7 March 2011).

McGray, H., Hammill, A., Bradley, R., Schipper, E. L. and Parry, J.-E. (2007) *Weathering the Storm: Options for Framing Adaptation and Development*, Washington: World Resources Institute. Available: http://www.wri.org/publication/weathering-the-storm (21 May 2012).

McClymont, D. P. and Myers, E. (2012) 'Community-based participatory process—Climate change and health adaptation program for Northern First Nations and Inuit in Canada,' *International Journal of Circumpolar Health*, vol. 71.

Meyer, W. B., Butzer, K. W., Downing, T. E., Turner II, B. L., Wenzel, G. W. and Wescoat, J. L. (1998) 'Reasoning by analogy,' in S. Rayner and E. L. Malone (eds.) *Human Choice and Climate Change. The Tools for Policy Analysis*, vol. 3. Washington, DC: Battelle Press, pp. 217–289.

Ministério para Coordenação da Acção Ambiental (MICOA) (2007) Direcção Nacional de Gestão Ambiental. *Programa de Acção Nacional para a Adaptação às Mudanças Climáticas (NAPA)*. Available: http://www.convambientais.gov.mz/index.php?option=com_docman&task=cat_view&Itemid=&gid=92&orderby=dmdate_published&ascdesc=DESC (28 September 2012).

Mirza, M. M. Q. (2003) 'Climate change and extreme weather events: can developing countries adapt?,' *Climate Policy*, vol. 3, no. 3, pp. 233–248.

Moraes, A. and Perkins, P. E. (2009) 'Etica, Genero e Classe Social na Politica Participativa de Agua, in G. A. dos Santos (ed.) *Etica, Pesquisa e Políticas Publicas*, São Paulo, Brazil: Editora Rubio, pp. 107–122.

Morrissey, J. (2006) *Livelihoods at Risk: The case of the Mphanda Nkuwa Dam*. Community Risk Assessment for the Proposed Mphanda Nkuwa Hydroelectric Dam. Executive Summary. Prepared for Justiça Ambiental. Available: http://www.open.ac.uk/technology/mozambique/pics/d97778.pdf (15 August 2012).

Muller, M. (2007) 'Adapting to climate change: water management for urban resilience,' *Environment and Urbanization*, International Institute for Environment and Development, vol. 19, no. 1, pp. 99–113. Available: http://eau.sagepub.com/cgi/reprint/19/1/99 DOI: 10.1177/0956247807076726.

Nakashima, D., Galloway McLean, K., Thulstrup, H., Ramos Castillo, A. and Rubis, J. T. (2012) *Weathering Uncertainty: Traditional knowledge for climate change assessment and adaptation*, Paris and Darwin: UNESCO and UNU.

Naidoo, P. (2009) *The making of 'The Poor' in post-apartheid South Africa*, Masters research thesis, University of KwaZulu-Natal School of Development Studies, Durban.

Neumayer, E. and Plümper, T. (2007) 'The gendered nature of natural disasters: the impact of catastrophic events on the gender gap in life expectancy, 1981–2002,' *Annals of the Association of American Geographers*, vol. 97, no. 3, pp. 551–566. DOI: 10.1111/j.1467–8306.2007.00563.x

Newell, P., Phillips, J. and Mulvaney, D. (2011) *Pursuing clean energy equitably*. Human Development Research Paper, United Nations Development Programme.

Ngwane, T. (2009) *Ideology and agency in protest politics*. Masters research thesis proposal, University of KwaZulu-Natal School of Development Studies, Durban.

Nojiyeza, S. and Galvin, M. (2011) 'Black People want to Experience the Flush: Expectations and experiences of urine diversion toilets in eThekwin,' Workshop presented at the University of KwaZulu-Natal School of Development Studies, 20 June 2011. Abstract available: http://ccs.ukzn.ac.za/default.asp?11,65,3,2329 (3 December 2012).

Notícias (2012) 'Sociedade civil debate hoje mudanças climáticas,' Maputo, Mozambique. Available: http://www.jornalnoticias.co.mz/pls/notimz2/getxml/pt/contentx/1504932 (28 September 2012).

Ostrom, E. (1990) *Governing the Commons: The Evolution of Institutions for Collective Action*, New York: Cambridge University Press.

Paavola, J., Adger, W. N. and Huq, S. (2006) 'Multifaceted justice in adaptation to climate change' in W. N. Adger *et al.* (eds.) *Fairness in Adaptation to Climate Change*, Cambridge, MA/London, UK: MIT Press, pp. 263–278.

Parry, M. L., Canziani, O. F., Palutikof, J. P., van der Linden, P. J. and Hanson, C. E. (eds.) (2007) *Contribution of Working Group II to the Fourth Assessment Report of the Intergovernmental Panel on Climate Change*. Cambridge, UK and New York: Cambridge University Press.

Pelling, M. (2003) *The Vulnerability of Cities: Natural Disasters and Social Resilience*. London: Earthscan.

Perkins, P. E. (2003) 'Public participation in ecological valuation: how policies can help it happen,' Paper presented at the Conference of the Canadian Society for Ecological Economics (CANSEE), Jasper, Alberta, October 16–19.

Perkins, P. E. (2004) 'Participation and watershed management: experiences from Brazil,' Paper presented at the Conference of the International Society for Ecological Economics (ISEE), Montreal, Canada, July 10–14.

Perkins, P. E. (2011) 'Social vulnerability and climate justice: community-based strategies for political engagement,' submitted December 2, 2011 for publication in A. Awotona (ed.) *A Global Survey of Sustainable Development in Areas of Social Vulnerability*.

Perkins, P. E. and Leary, A. L. T. (2012) 'Climate justice partnership linking universities and community organizations in Toronto, Durban, Maputo, and Nairobi,' in W. Leal (ed.) *Sustainable Development at Universities: New Horizons*. Frankfurt: Peter Lang Scientific Publishers, pp. 207–215.

Pettengell, C. (2010) 'Climate Change Adaptation: Enabling people living in poverty to adapt,' *Oxfam International Research Report*, Oxford: Oxfam International. Available: http://publications.oxfam.org.uk/results.asp?sf1=contributor&st1=Catherine%20Pettengell&TAG=&CID= (4 May 2011).

Pieterse, M. (2007) 'Eating Socioeconomic Rights: The Usefulness of Rights Talk in Alleviating Social Hardship Revisited,' *Human Rights Quarterly*, vol. 29, pp. 796–822.

Ramsay, L. (2009) *Climate Change for the People of South Durban*. Durban: South Durban Community Environmental Alliance, "Touching Tomorrow" series, edition 3.

Roberts, D. (2008) 'Thinking globally, acting locally – institutionalizing climate change at the local government level in Durban,' *South Africa Environment and Urbanisation*, vol. 20, no. 2, pp. 521–537.

Sagoff, M. (1998) 'Aggregation and deliberation in valuing environmental public goods: a look beyond contingent pricing,' *Ecological Economics*, vol. 24, pp. 213–230.

Sanderson, D. (2000) 'Cities, disasters and livelihoods,' *Risk Management: an International Journal*, vol.2, pp. 49–58.

Satterthwaite, D., Huq, S., Reid, H., Pelling, M. and Romero Lankao, P. (2007) *Adapting to climate change in urban areas: The possibilities and constraints in low- and middle-income nations*. International Institute for Environment and Development, Human Settlements Discussion Paper Series, Climate Change and Cities—1. Available: http://www.iied.org/HS/topics/accc.html and http://www.iied.org/pubs/display.php?o=10549IIED (14 July 2011).

Scott, J. (1992) *Domination and the Arts of Resistance*. New Haven: Yale University Press.

Sereda, B. L. and Meinhardt, H. R. (2005) 'Contamination of the Water Environment in Malaria Endemic Areas of KwaZulu Natal, South Africa by DDT and Its Metabolites,' *Bulletin of Environmental Contamination and Toxicology*, vol. 75, pp. 538–545.

Sister Watersheds, 2008. Project website: http://www.yorku.ca/siswater (11 October 2011).

Smit, B. and Pilifosova, O. (2003) 'From adaptation to adaptive capacity and vulnerability reduction,' in J. B. Smith, R. J. T. Klein and S. Huq (eds.) *Climate Change, Adaptive Capacity and Development*, London: Imperial College Press.

Smit, B. and Wandel, J. (2006) 'Adaptation, adaptive capacity and vulnerability,' *Global Environmental Change*, vol. 16, pp. 282–292.

Speer, P. W. and Hughey, J. (2008) 'Community organizing: an ecological route to empowerment and power,' in J. Defilippis, and S. Saegert (eds.) *The Community Development Reader*, New York: Routledge, pp. 204–213.

Swyngedouw, E. (2008) 'Retooling the Washington Consensus: The contradictions of H_2O under neo-liberalism and the tyranny of participatory governance,' Paper presented at the Centre for Civil Society, University of KwaZulu-Natal, Durban, 3 July.

Taylor, V., Schulze, R. E. and Jewitt, G. P. W. (2003) 'Application of the Indicators of Hydrological Alteration method to the Mkomazi River, KwaZulu-Natal, South Africa' *African Journal of Aquatic Science*, vol. 28, no. 1, pp. 1–11.

Terry, G. (ed.) (2009) *Climate Change and Gender Justice*, Rugby, UK: Practical Action Publishing in Association with Oxfam GB.

Tompkins, E. L. and Adger, W. N. (2004) 'Does adaptive management of natural resources enhance resilience to climate change?' *Ecology and Society*, vol. 9, no. 10. Available: http://www.ecologyandsociety.org/vol9/iss2/art10 (14 September 2012).

Toulmin, C. (2010) *Climate Change in Africa*. London and New York: Zed Books.

Toth, F. L. (1999) *Fair Weather? Equity Concerns in Climate Change*. London: Earthscan.

Turton, A. R. (1999) 'Water scarcity and social adaptive capacity: towards an understanding of the social dynamics of water demand management in developing countries,' MEWREW Occasional Paper No. 9, University of London School of African Studies, London.

UN Development Programme (2010) *Community Based Adaptation to Climate Change*. Available: http://www.undp.org/climatechange/library_adaptation.shtml (21 July 2012).

UNEP (United Nations Environment Programme) (2008) UNEP Sourcebook: Integrating Adaptation to Climate Change into UNEP Programming. Nairobi, Kenya, 8–9 April.

UN Women Watch (2009) *Fact Sheet: Women, Gender Equality and Climate Change*. Available: http://www.un.org/womenwatch/feature/climate_change/downloads/Women_and_Climate_Change_Factsheet.pdf (September 2012).

United Nation Human Settlements Program (UN Habitat) (2010) *Climate Change Assessment for Maputo, Mozambique*. Cities and Climate Change Initiative. Nairobi, Kenya. Available: http://www.unhabitat.org/pmss/getElectronicVersion.aspx?nr=2977&alt=1 (5 May 2011).

van den Hove, S. (2000) 'Participatory approaches to environmental policy-making: the European commission climate policy process as a case study,' *Ecological Economics*, vol. 33, pp. 457–472.

WWF—World Wide Fund For Nature (2006) *Climate Change Impacts on East Africa: A review of the scientific literature*. WWF, Gland, Switzerland. Available: http://www.wwf.org.uk/filelibrary/pdf/e_africa_climatechange_report.pdf (31 May 2011).

Walker, B., Carpenter, S. R., Anderies, J., Abel, N., Cumming, G. S., Janssen, M., Lebel, L., Norberg, J., Peterson, G. D. and Pritchard, R. (2002) 'Resilience management in social–ecological systems: a working hypothesis for a participatory approach,' *Conservation Ecology*, vol. 6. Available: http://www.ecologyandsociety.org/vol6/iss1/art14/print.pdf (28 September 2012).

WaterAid (n.d.) *Mozambique*. Available: http://www.wateraidamerica.org/includes/documents/cm_docs/2008/m/mozambique_a4.pdf (28 September 2012).

Wilson, M. A. and Howarth, R. B. (2002) 'Discourse-based valuation of ecosystem services: establishing fair outcomes through group deliberation,' *Ecological Economics*, vol. 41, pp. 431–443.

Winters, P., Murgai, R., Sadoulet, E., de Janvry, A. and Frisvold, G. (1998) 'Economic and welfare impacts of climate change on developing countries,' *Environmental and Resource Economics*, vol. 12, pp. 1–24.

Whitfield, A. K. and Taylor, R. H. (2009) 'Case Studies and Reviews: A review of the importance of freshwater inflow to the future conservation of Lake St Lucia,' *Aquatic Conservation: Marine and Freshwater Ecosystems*, vol. 19, pp. 838–848.

Yohe, G. and Tol, R. (2002) 'Indicators for social and economic coping capacity—moving toward a working definition of adaptive capacity,' *Global Environmental Change*, vol. 12, pp. 25–40.

Index

Page numbers in *italic* refer to Figures, **bold** numbers indicate Boxes

For Product Safety Concerns and Information please contact our EU representative GPSR@taylorandfrancis.com
Taylor & Francis Verlag GmbH, Kaufingerstraße 24, 80331 München, Germany

www.ingramcontent.com/pod-product-compliance
Ingram Content Group UK Ltd.
Pitfield, Milton Keynes, MK11 3LW, UK
UKHW020130200926
475702UK00005B/678
9780415822039